Science Super Sleuths

by Pam Walker and Elaine Wood

illustrated by Rex Schneider

cover by Jeff Van Kanegan

Publisher
Instructional Fair • TS Denison
Grand Rapids, Michigan 49544

ISBN: 1-56822-843-0

Science Super Sleuths

a division of Tribune Education

2400 Turner Avenue NW

Grand Rapids, Michigan 49544

Unit 1—Life Science

Unit 2—Physical Science

Unit 3—Earth Science

To the Teacher

To peak their interest, science questions can be posed to students as mysteries that need to be solved. In reality, scientists actually see the problems they want to solve as mystery questions. In a sense, scientists are detectives who are always looking for clues and planning a strategy for finding their answers. Oftentimes, both scientists and detectives begin their search with very little information. However, by thinking through problems, working with other people, and listening to new ideas, they can reach their goals.

Each activity in *Science Super Sleuths* presents students with a Science Sleuth Question, a mystery for them to solve. To help them get started, students are provided with Background Information that tells them all they need to know to devise a plan for solving their mystery. Pre-Lab Questions, which follow the background information, check students' reading comprehension.

Working in groups of two or three, students can develop and record their experimental plan, select and list the equipment they need, write their procedure, and conduct their experiment with little or no teacher intervention. They are encouraged to create Data Tables and graphs to organize their findings. Conclusion Questions help the teacher find out whether or not students understand their experimental results.

All of the labs use safe and inexpensive materials. In each lab, we suggest that the teacher make two kinds of materials available to students: those which the students must have to complete the lab and some extraneous materials that are unrelated to the lab. This arrangement forces students to make some decisions about their plan before they can get started.

A generic Suggested Evaluation Rubric is provided below to help grade student work. This rubric may need slight modifications for some of the labs, but can generally be used as is.

On the Teacher Information page of each activity, Suggested Experimental Setup is provided. This information saves teachers planning time: teachers can check the Suggested Experimental Setup and compare it to what students have developed. Any suggested Experimental Setup can also be used as a prescribed procedure, if desired.

Suggested Evaluation Rubric

Criteria	Points Possible	Points Earned
Pre-Lab Questions answered	15	________
Summary of Experimental Plan	15	________
Materials listed	15	________
Procedure described	10	________
Data Table complete	15	________
Experiment conducted	15	________
Conclusion Questions answered	15	________
Total	100	________

Unit 1

Life Science

Mummified

Objective: Students will design and conduct an experiment in which they mummify a piece of fruit.

Time Required
Day 1: 55 minutes
Days 2-13: 10 minutes/day
Day 14: 44 minutes
Day 15: 10 minutes

Teaching Strategies

In this full inquiry, students are provided with background information on food preservation and mummification so that they can develop a strategy for mummifying fruit.

Instead of telling students which materials they need, provide them with a variety of items and let them select what they consider to be most appropriate. Some lab material suggestions include sugar, salt, several types of fruit, paper cups and plates, plastic spoons, detergent, water, milk, aluminum foil, scales, and graduated cylinders.

Suggested Experimental Setup

Have students slice nine pieces of fruit and then weigh them. On each of three paper plates write the following: Group A, Group B, and Group C. Place fruit slices 1–3 on the plate labeled Group A, slices 4–6 on the plate labeled Group B, and slices 7–9 on the one labeled Group C. Record the appearance of the fruit in each group.

Fruit in Group A serves as the control. Cover fruit in Group B with salt. Cover fruit in Group C with sugar. Observe the fruit daily and record its appearance and weight. (Shake off as much sugar and salt as possible before weighing.) During the experiment, the fruit covered in salt or sugar dehydrates significantly and is well on its way to mummification.

Figure 1. In this experiment, place three slices of fruit on paper plates that are labeled Group A, Group B, and Group C. Cover the fruit in Group B with salt and the fruit in Group C with sugar.

Background Information

Mummified

??????? Science Sleuth Question ?????

What causes mummification?

You are living in a wild place, far from grocery stores and refrigerators. Finding food is your most important activity. This is a great day because in your food search you locate an apple tree that is loaded with red apples. You pick all of the apples you can carry and head back to camp. On the way, a check on your rabbit trap reveals that you have made a catch. You quickly kill the rabbit, tie it to your waist with a piece of vine, and continue on your way.

Satisfied that you have enough food to last for several days, you look forward to several meals of rabbit meat and apples. At the camp, you clean and cook your rabbit, then sit back and enjoy it with those delicious apples. You wrap the leftover food in leaves and hide the bundle under a rock.

Figure 1. Some apples and a freshly killed rabbit cooked over the campfire provide a tasty meal.

The next day, you remove your food bundle from its hiding place and find that the rabbit smells and tastes terrible. The apples are still fine, so you eat a few of them and hide the food one more time. On the third day, you remove your rabbit to find that it is covered with mold and has begun to attract insects. You throw the rabbit meat away, wondering what happened to your food. The apples are still edible.

Unfortunately, your rabbit meat decayed or spoiled. Dead animals and plants spoil because bacteria and fungi feed on them. In the feeding process, these organisms change the dead plants and animals into simple, easy-to-digest compounds. Some of these compounds smell bad. Most of them taste terrible, too.

Long ago, people learned to preserve food so they could store it for long periods of time. For many years, the most common methods of food preservation were drying, salting, and cooling.

a. When food is dried, all of the water is removed. Bacteria and fungi, like all living things, must have water to live. Therefore, they cannot live on dried food. Some dehydrated foods are cereals and fruits.

b. Large amounts of salt or sugar added to food cause all of the water to leave the food. Preserving with salt and sugar works much like drying food. Some meats are cured by salting. Sugar is used to preserve jams and jellies.

c. Refrigeration and freezing slow the growth of bacteria and fungi. Cooling food is a relatively new way to preserve it.

When people die, their bodies decay unless they are preserved in some way. Today, if someone dies mysteriously, the body may be kept in a refrigerator until an investigation can be conducted. A fairly new, long-term method of body preservation uses a chemical that changes or cross-links the tissue. Cross-linked tissue cannot be broken down by bacteria and fungi. You may have dissected frogs preserved in such chemicals.

The Egyptian mummies are some of the most fascinating examples of preserved humans. Even though the Egyptian culture existed thousands of years ago, the bodies of some of these people are still intact today. In ancient Egypt, the body of a dead person was bathed in a variety of salts and spices, then wrapped in strips of cloth. In many ways, mummification is very similar to salting or sugaring food. Scientists are still learning about the ways Egyptians preserved their dead.

Pre-Lab Questions

1. What causes dead plants and animals to decay?

____

____

____

2. Why must food be preserved?

____

____

____

3. List and explain three ways to slow the decay of dead plants and animals.

____

____

____

4. Compare the methods used by Egyptians to preserve their dead to the method we use to preserve humans today.

____

____

____

Name ______________________

Mummified

***Objective:** Students will design and conduct an experiment in which they mummify a piece of fruit.*

Summary of Your Experimental Plan: ______________________

____

____

____

____

____

____

Materials Needed: ______________________

____

____

Procedure: ______________________

____

____

____

____

____

____

____

____

____

____

____

____

Data Table: (if needed)

Results: ____

____

____

____

____

____

____

Conclusion Questions

1. What type of fruit did you attempt to mummify? ____________________

2. At the end of the experiment, how did you know whether the fruit was mummified?

____

____

3. Why was this experiment conducted for 14 days instead of just one day?

____

____

4. How are the techniques of mummifying food similar to the ways Egyptians mummified their dead?

____

____

"Lite" Eaters

Objective: Students will compare the volume and mass of water in three light or low-fat brands of margarine.

Time Required
Day 1: 55 minutes

Teaching Strategies

In this partial inquiry, students are provided with Background Information that helps them develop a strategy for comparing the volume and mass of water in three margarine samples.

Students will most likely need a hot plate or Bunsen burner to melt their butter samples. They also need containers for the butter, such as test tubes, beakers, or jars. To compare the volume or mass of water in each type of margarine, students can use graduated cylinders or scales. Other materials that you might make available in the lab include hot pads, beakers, tongs, plastic spoons or knives, paper towels, paper plates, and vegetable oil. Be certain to include some lab materials that are completely irrelevant to the lab so that students will learn to think and plan before they gather their equipment.

Suggested Experimental Setup

Label the three brands of light margarine as A, B, and C. Have students label three small beakers as A, B, and C. In each beaker, place 50 grams of the appropriate sample of margarine.

Place the three beakers of margarine in a hot water bath so that the margarine melts. Decant the oil from beaker A; then measure the water left in that beaker. Record this information in a Data Table. Repeat this procedure with beakers B and C.

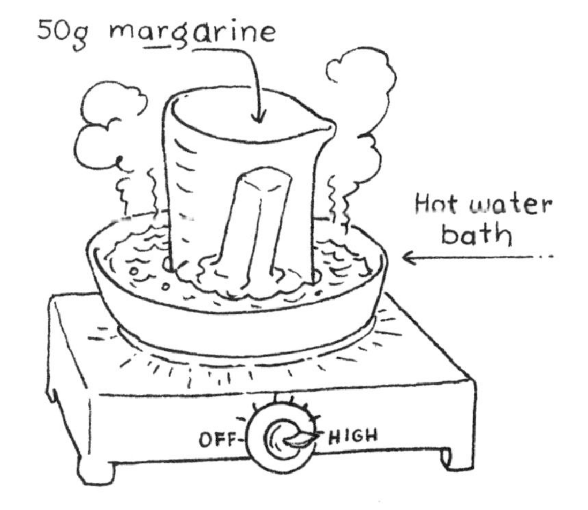

Figure 1. Place 50 grams of margarine in each beaker. Then warm the beakers in a hot water bath.

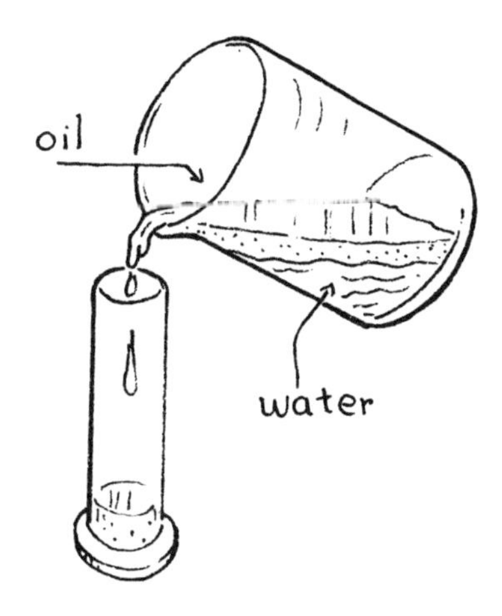

Figure 2. Decant oil off melted margarine.

Background Information

"Lite" Eaters

? ? ? ? ? **Science Sleuth Question** ? ? ? ? ?

Which brand of light margarine contains the most water?

As a group, Americans have more fat in their diets than people from any other country. Although nutritionists suggest that we limit fat intake to 25% or 30% of our total calories each day, many people's diets contain up to 50% fat. Fat in our diets has been linked to heart disease, hardening of the arteries, cancer, and obesity. However, we are beginning to learn to eat smarter. And as a result, the food industry is giving us some choices in the supermarket.

While grocery shopping, consumers are bombarded with choices. Many products are modified to be "light," "lite," and "low" in fat or other ingredients. These labels suggest that the food is good for you because it has less of some offending ingredient. But if it has less of one ingredient, it must have more of another. Do you know what is replacing fat in your food? The answer seems simple: read about the ingredients on labels. However, the labels do not clearly explain the amount of each ingredient in the product.

Many foods are made "light" by the addition of water. Water has no fat or calories, so products that replace fat with water are indeed "lighter." However, water may not be what you want to buy. And strange as it seems, addition of water to a food sometimes makes it more expensive.

Fats and oils are similar compounds. They are both classified as *lipids*, naturally occurring substances that do not dissolve in water. Fats are usually solids at room temperature, while oils are liquids. Because water and oil will not mix, it is easy to tell whether or not an oil-based product has had water added to it. By liquefying the product and pouring it into a tall glass, you can see if it separates into two layers. If it does, it probably contains water.

Figure 1. Foods that contain lipids include margarine, butter, milk, eggs, meats, poultry, and fish.

Name ______________________________

Pre-Lab Questions

1. What do the words *lite, light,* and *low* indicate about food? ____________________
____
____

2. How can you tell whether or not a bottle of liquid is all oil, or a mixture of oil and water? ____
____
____

3. Imagine that you are a margarine manufacturer. Which material do you think costs you the most to buy: water or oil? ____________________________________
____
____

4. How much of your daily total calories should be fat? __________________________
____
____

5. What are lipids? ____
____
____

Name ______________________

"Lite" Eaters

Objective: *Students will compare the volume and mass of water in three light or low-fat brands of margarine.*

Summary of Your Experimental Plan: ______________________

Materials Needed: ______________________

Procedure: ______________________

Data Table:

Margarine Samples	Volume of water in 10 grams of margarine	Mass of water in 10 grams of margarine
Sample #1		
Sample #2		
Sample #3		

Results: ____

____

____

____

____

____

____

Conclusion Questions

1. What are the names of the three brands of margarine you used in this lab?

   ____

   ____

2. What is the price per pound of each brand of margarine?

   ____

3. Judging from the names of the margarines, which did you expect to contain the least amount of fat (and, therefore, the smallest number of calories)? ____________

   ____

   ____

4. Which margarine costs the most: the one with the least amount of water or the one with the most water? ______________________________

5. If "light" or "low fat" margarine were not available in a grocery store, how could you make your regular margarine "light"? ______________________________

   ____

Bee Stings and Beef Steaks

Objective: Students will design and conduct an experiment to determine which substance contains the most active protein-destroying enzymes: fresh pineapple juice or meat tenderizer.

Time Required
Day 1: 55 minutes
Days 2-5: 10 minutes
Day 6: 30 minutes

Teaching Strategies

In this full inquiry, students are given some information on proteins and the activity of enzymes. Enzymes can be found all through the body. For example, saliva contains an enzyme that breaks down starch. However, meat tenderizer, fresh pineapple juice, and contact lens cleaner may be safer sources of enzymes in the classroom. The enzymes in these three products are classified as *proteases* because they digest or break down proteins.

Provide students with a variety of lab materials, including gelatin cubes. Gelatin cubes are the simplest proteins with which students can work. Before lab, prepare some clear gelatin; then cut it into cubes for students. If you or the students want to quantify the amount of protein breakdown that occurs, have them weigh/measure the cubes before, during, and after the lab. Some suggestions for other lab materials include scales, paper plates or petri dishes, droppers or pipettes, teaspoons, and rulers.

Suggested Experimental Setup

Label three paper cups as A, B, and C. In cup A, stir a few teaspoons of water into a little meat tenderizer. Pour a little fresh pineapple juice into cup B. Pour a little water into cup C.

Place three cubes of gelatin on three paper plates. Label the plates as A, B, and C. On the gelatin cubes in plate A, add a few drops of solution from cup A. Treat the other two plates in a similar fashion. Observe the cups every five minutes on Day 1. Record any changes that you observe. Refrigerate the gelatin cubes overnight and examine them daily through the sixth day.

Figure 1. suggested experimental setup showing three paper plates, each containing three gelatin cubes

Background Information

Bee Stings and Beef Steaks

? ? ? ? ? Science Sleuth Question ? ? ? ? ?

Which is the best source of a protein-destroying enzyme: fresh pineapple juice or meat tenderizer?

It could be said that proteins make you who you are. They form most of your body parts, as well as the glue that keeps those parts together. A special group of proteins, called *enzymes*, has an essential role in regulating the rates of chemical reactions. Without enzymes, you would not be alive.

Whether you are aware of it or not, you have experienced the activity of proteins. Have you ever been stung by a bee or wasp? The venom of these animals is made up of proteins that make your skin burn or itch. The pain eventually goes away because the protein breaks down into simpler compounds. The sooner the protein-based venom is gone, the sooner the pain stops.

The next time you get a sting, treat it with a little meat tenderizer. Meat tenderizer contains a type of enzyme that can break down proteins. We generally buy meat tenderizer to sprinkle on meat before we cook it. Some meat is tough because of proteins that connect the muscle fibers together. Meat tenderizer breaks down these connective proteins, thereby separating the muscle fibers. That makes the meat easier to chew.

Meat tenderizer can also be used to destroy other proteins. If you rub a little damp meat tenderizer on a fresh bee sting, it breaks down the protein in the venom. Other familiar items that contain protein-destroying enzymes include fresh pineapple juice, some laundry detergents, and contact lens cleaner.

Proteins can be found in many materials, such as meat, gelatin, hair, spider webs, and silk. The enzyme in meat tenderizer, pineapple juice, and contact lens cleaner will speed up the breakdown of any of these proteins.

Figure 1. The venom of bees and wasps is made of proteins.

Pre-Lab Questions

1. What are two jobs performed by proteins in the body? ____________________

   ____

2. List some examples of proteins. ____________________

   ____

3. Where can you find some enzymes that break down proteins? ____________________

   ____

Name ______________________

Bee Stings and Beef Steaks

Objective: *Students will design and conduct an experiment to determine which substance contains the most active protein-destroying enzymes: fresh pineapple juice or meat tenderizer.*

Summary of Your Experimental Plan: ______________________

____

____

____

____

____

____

Materials Needed: ______________________

____

____

Procedure: ______________________

____

____

____

____

____

____

____

____

____

____

____

____

Data Table: (if needed)

Results: ______________________________

____

____

____

____

____

____

Conclusion Questions

1. What protein material did you expose to enzymes? ______________________________

   ____

2. What changes in your protein sample indicated that the protein was being destroyed by the enzymes? ______________________________

3. Which contained the most active enzymes: fresh pineapple juice or meat tenderizer?

   ____

4. Why did we conduct this experiment over a period of several days? ______________________________

   ____

5. Why do you think that contact lens cleaner contains a protein-digesting enzyme?

   ____

6. What is the function of protein-digesting enzymes in detergents? ______________________________

   ____

7. Where in your body would you expect to find protein-digesting enzymes? ______________________________

   ____

Keep Me Warm and Neutral

Objective: Students will design and conduct an experiment to determine how changes in pH and/or temperature affect the activity of an enzyme.

Time Required
Day 1: 55 minutes
Day 2: 30 minutes

Teaching Strategies The activity of enzymes is explained to students in the Background Information. In this lab, use the same proteins and enzymes that you used in Bee Stings and Beef Steaks. Students can vary the pH or the temperature of the enzymes and then perform the same kind of experiment that they used in the previous lesson. Enzymes function best at temperatures between 50 and 98 degrees Fahrenheit and at a pH near 7.

Discuss experimental control groups with students. Ideally, students should establish a control setup in which the pH and temperature of enzymes have not been altered. However, the previous lab can serve as your control group.

Provide students with a variety of lab materials, including gelatin cubes. These are the simplest proteins with which they can work. Before the lab, prepare some clear gelatin; then cut it into cubes for students. If you or the students want to quantify the amount of protein breakdown that occurs, have them weigh/measure the cubes before, during, and after the lab. Some suggestions for other lab materials include scales, paper plates or petri dishes, droppers or pipettes, rulers, hot plates, fresh pineapple juice, contact lens cleaner, meat tenderizer, vinegar, household ammonia, baking soda solution, spoons or stirring rods, pH paper, lemon juice, laundry detergent, beakers, and aluminum foil.

Suggested Experimental Setup Use gelatin cubes as proteins to be tested. Place cubes into three groups and label them as Group 1, Group 2, and Group 3. Add a few drops of the enzyme (from meat tenderizer, fresh pineapple juice, or contact lens cleaner) to Group 1 cubes. Warm a sample of this enzyme in a hot water bath, allow it to cool, and then add it to gelatin cubes in Group 2. To another sample of enzyme, add a few drops of an acid (vinegar or lemon juice) or a base (ammonia or baking soda solution). Check the pH; then place a few drops of this acidic (or alkaline) enzyme on Group 3 gelatin cubes.

Observe the effects of the enzymes on all three groups of gelatin cubes. Over a short period of time (one hour to one day), the Group 1 cubes will be digested. However, heating and changing the pH affected the enzymes used in Groups 2 and 3, so those gelatin cubes will not be changed. Remember to refrigerate the gelatin before and after class.

Background Information

Keep Me Warm and Neutral

? ? ? ? ? Science Sleuth Question ? ? ? ? ?

What factors affect the activity of enzymes?

Chemical reactions occur in your body all the time. These reactions would occur too slowly to maintain life if they were not aided by enzymes. *Enzymes* are large protein molecules that trigger and control the rates of chemical reactions in living things. The materials that are changed by a chemical reaction are called *substrates*. Enzymes attach to substrates and help change them.

For example, in many chemical reactions substrates are taken apart to produce new substances. This is the kind of reaction that occurs when a protein is broken down into simpler parts. An enzyme helps this reaction occur by holding the protein in position so that it can be changed.

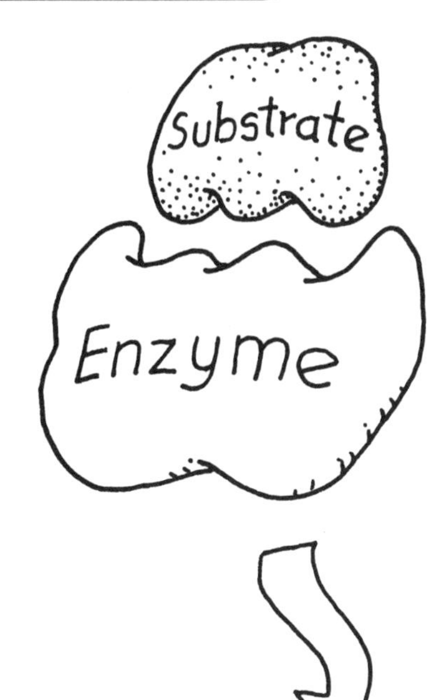

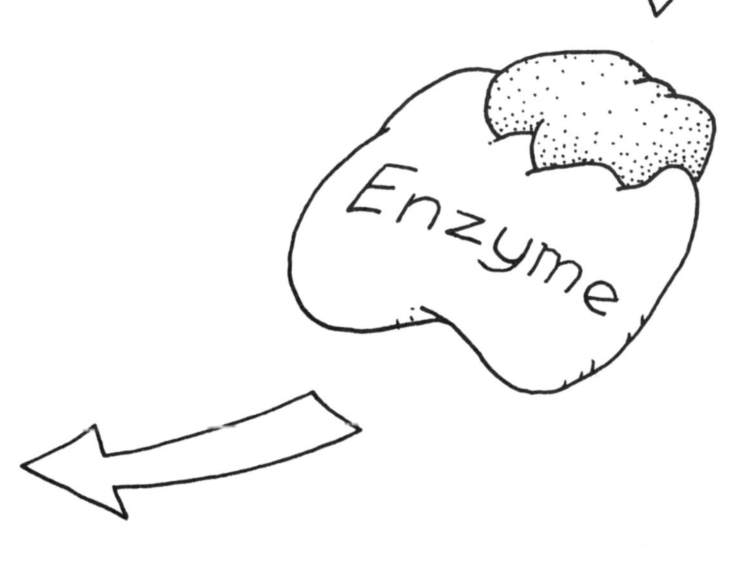

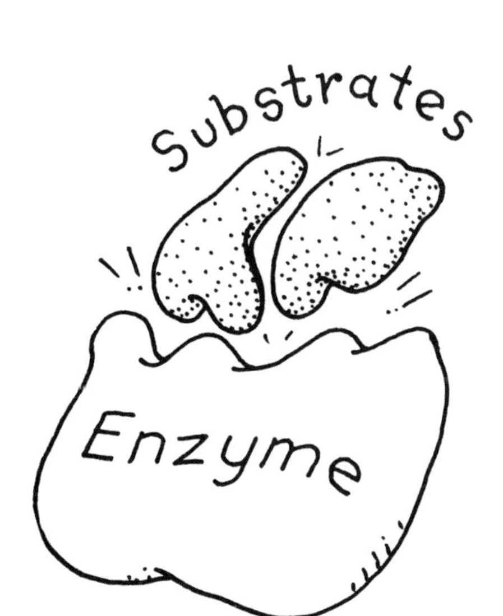

Figure 1. Some enzymes help break substrates down into simpler substances.

Name ______________________

During other chemical reactions, substrates are joined together. For example, in your body small protein parts are joined together to make a protein. This is the kind of reaction that occurs when your body uses digested food to make new tissues. New tissue is created when you grow or when an injury heals. An enzyme helps this reaction occur by positioning the substrate pieces so that they can be united.

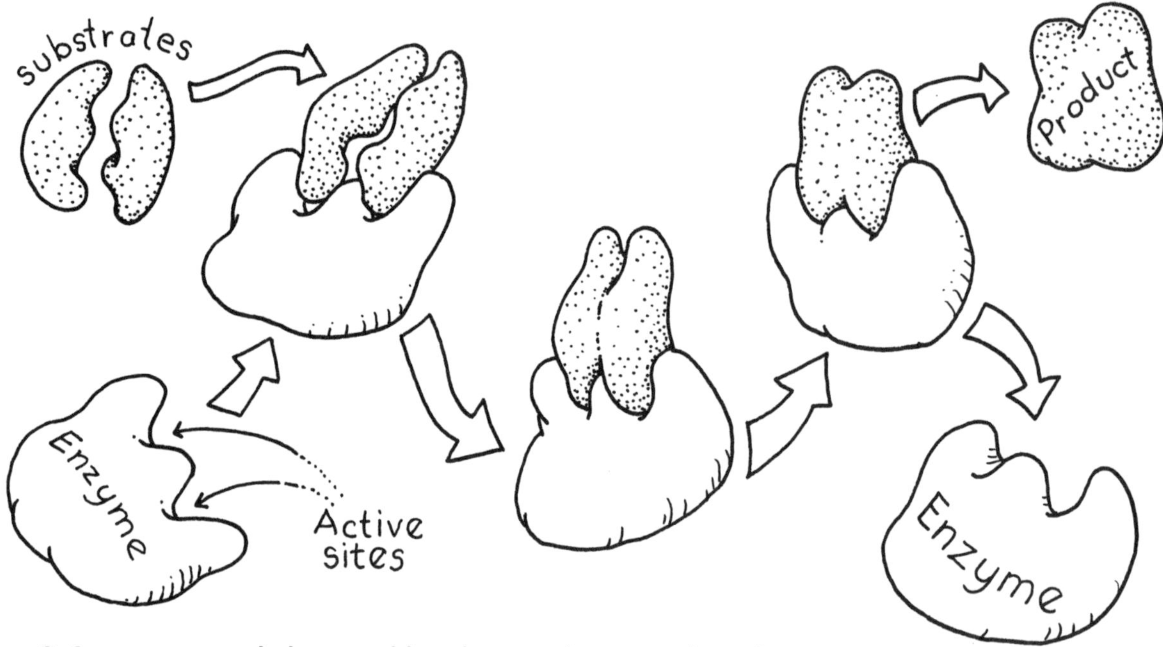

Figure 2. Some enzymes help assemble substrates into complex substances.

Enzymes are delicate substances that are easily destroyed. Two of the most common ways of destroying enzymes are by heating them or by changing the pH of their environment. The term *pH* indicates how acidic or basic a material is. Enzymes work best at a neutral pH—one that is neither acidic or basic. That is why it is important that your body maintain a neutral pH.

Your body also maintains a steady temperature to protect its enzymes. Heat distorts the shape of enzymes. Once their shape is changed, enzymes are no longer able to attach to their substrates. This prevents them from participating in chemical reactions.

Pre-Lab Questions

1. What factors affect the activity of enzymes? ______________________

   ____

2. How do enzymes function? ______________________

   ____

3. What are substrates? ______________________

4. Give one example of a chemical reaction that depends on an enzyme. ______________________

   ____

Name ____________________

Keep Me Warm and Neutral

Objective: *Students will design and conduct an experiment to determine how changes in pH and/or temperature affect the activity of an enzyme.*

Summary of Your Experimental Plan: ____________________

Materials Needed: ____________________

Procedure: ____________________

Data Table: (if needed)

Results: ______________________________

____

____

____

____

____

____

Conclusion Questions

1. How did you increase the enzyme's temperature? ______________
   How did this temperature change affect the enzyme's activity? ______________
   ____
2. How did you alter the pH of the enzyme? ______________
   How did this change affect the enzyme's activity? ______________
   ____
3. Some contact lens cleaners contain enzymes. Predict the pH of these cleaners. ______
   ____
4. Factors that change the shape of an enzyme affect its activity. Give two examples of such factors. ______________
   ____
5. Himalayan cats have dark-colored paws, ears, and tails. The chemical reaction that produces the dark pigment in these body parts requires an enzyme that is only active at certain temperatures. Which of the following choices best describes the temperature at which this enzyme is active:
   a. higher than body temperature
   b. same as body temperature
   c. cooler than body temperature
6. If the protein-digesting enzymes in your stomach were to change shape, what effect would that have on your body? ______________
   ____

Plants to Dye For

Objective: Students will choose a plant and use it to make a dye. They will then dye a piece of fabric.

Time Required
Day 1: 45 minutes
Day 2: 45 minutes
Day 3: 30 minutes

Teaching Strategies On Day 1, take students outdoors to collect the plants to use in the lab. Remind students to collect enough plant material to make a significant amount of dye. Mature plants yield larger quantities of dyes than young ones. Some suggestions of plant parts and the colors they yield follow:

a. Barks—Apple (*Prunus malus* or *Malus sylvestris*) yellow
Birch *(Betula lutea)* dark yellow to yellow brown
White hickory *(Carya tomentosa* or *C. alba)* yellow brown
Norway maple *(Acer platanoides)* rose to tan
Oak *(Quercus)* gold
Walnut *(Juglans nigra)* black
Willow *(Salix nigra)* black

b. Leaves—Birch *(Betula lutea)* yellow
Privet *(Ligustrum)* gold
Mountain laurel *(Kalmia latifolia)* yellow-tan
Tulip tree *(Liriodendron tulipifera)* gold
Lombardy poplar *(Populus nigra italica)* brass

c. Stalk and leaves—Broom sedge *(Andropogan virginicus)* green

d. Fruits—Juniper berries *(Juniperus)* khaki
Onion skins *(Allium cepa)* burnt orange

e. Fruit hulls—Butternut hulls *(Juglans cinerea)* brown
Hickory nuts *(Carya laciniosa* or *Hicoria laciniosa)* brown
Pecan hulls *(Carya illinoensis* or *Hicoria pecan)* brown
Green walnut hulls *(Juglans nigra)* black

f. Flowers—Camomile *(Anthemis tinctoria)* yellow
Dahlia *(Dahlia)* orange
Goldenrod *(Solidago)* yellow
Marigold *(Tagetes)* dark yellow
Sun flowers *(Helianthus annuus)* yellow
Zinnias *(Zinnia)* yellow to green

g. Lichens—brown, yellow, red, or purple (depending on the species)

If you wish to save time, collect these plants for your students before class. Day 3, when students dye a piece of fabric, is optional. You can even encourage tie-dying if time allows.

The best dyebaths use soft water. You will need 4 to 4½ gallons of dyebath per pound of fabric. After dying, rinse fabrics in water that is the same temperature as the dyebath. When squeezing out excess water and dye, do not twist or ring.

If you would like to make the dye samples color safe, immerse them in a mordant the day before dying. Try one of the following (recipes are for 1 pound of dry fabric):
Alum mordant (wool): 4 ounces alum (aluminum potassium sulfate)
1 ounce cream of tartar
4 to 4½ gallons cold water
Warm fabric slowly in mordant. Let stand overnight. Squeeze out excess moisture and allow to dry.

Chrome mordant: ½ ounce of potassium dichromate
Dissolve potassium dichromate in 4 to 4½ gallons of cold soft water. Follow alum mordant instructions.

Alum mordant: 4 ounces alum
1 ounce laundry detergent
Dissolve alum and laundry detergent in 4 to 4½ gallons cold water. Follow wool alum mordant instructions.

In this full inquiry, students are provided with Background Information that gives them some general knowledge about how dyes can be extracted from plants. Students can use this information to formulate a plan for making dye.

Instead of telling students what materials they need, provide them with the following items and let them select what they feel is appropriate: hot plates or Bunsen burners, water, large beakers or pans, hammers, spoons, string, cheesecloth or sieves, samples of white or unbleached fabrics or yarns (cotton, wool, silk, and/or linen—do not use synthetic fabrics). It is not necessary for students to mordant fabric before dying unless they plan to wash or wear their dye samples.

Suggested Experimental Setup

Have students chop, cut, or mash their plant pieces, and then boil them in water. If boiling time is limited, let the plants sit in their containers of water overnight. Strain the plant material out of the water with sieves or cheesecloth. Have students wear aprons and goggles.

Background Information

Plants to Dye For

? ? ? ? ? Science Sleuth Question ? ? ? ? ?

What plants can be used to make fabric dyes?

Check out the colors of your clothes and those of your classmates. Color is an important part of fashion. The colors found in clothing are the results of dyes. Most of the dyes in your clothes are classified as *synthetic* because they were man-made. Synthetic dyes give a wide range of colors. However, at one time the colors produced by natural dyes were the only ones available.

Whether a dye is natural or synthetic, it must be dissolved in a solvent before it can work. The dye and its solvent create a dyebath. When fabric is placed in a dyebath, its fibers absorb the molecules of dye. These dye molecules give the fabric its desired color.

Fabrics that are dyed vary in their color fastness, or ability to hold the color. Most fabrics are wash fast, which means that they will not fade during normal laundering. To improve color fastness, fabrics can be immersed in mordants before they are dyed. Mordants are molecules that combine with the dye and fix the dye permanently in the fabric. Mordants include tannic acid and compounds of aluminum, chromium, tin, iron, and copper.

Many natural dyes are extracted from plants and animals. Two widely used animal dyes are red cochineal and Tyrian purple. Red cochineal comes from dried bodies of a certain insect. Tyrian purple is derived from the shell of the marine mollusk. This mollusk has always been rare. In the past, only the wealthy could afford this dye, so it was nicknamed "royal purple."

Hundreds of plants provide dyes. For example, an orange dye is made from dried stigmas of a small flower called the saffron crocus. Saffron is often used to dye wool and silk. A red dye can be made from the roots of the madder plant. A dark blue dye, indigo, is made from the indigo plant. It is still used commercially to dye blue jeans.

Most plant dyes are made by collecting, crushing, and boiling the desired plant parts in water. Plant parts are filtered out of the dye, which is reheated. Fabric is added to the boiling water and allowed to steep until the desired color is reached.

Figure 1. Saffron crocus

Pre-Lab Questions

1. What are natural dyes? ____________________

2. What are some sources of plant dyes? ____________________

3. Why do commercial dyers use mordants? ____________________

Name ______________________

Plants to Dye For

Objective: *Students will choose a plant and use it to make a dye. They will then dye a piece of fabric.*

Summary of Your Plan to Make a Dye: ______________________

Materials Needed: ______________________

Procedure: ______________________

Data Table:

Fabric Samples	Dye Absorbed Evenly in Fabric (Yes/No)	Color of Dye Is Deep and Rich (Yes/No)
Fabric Sample #1		
Fabric Sample #2		
Fabric Sample #3		

Results:____

____

____

____

____

____

____

Conclusion Questions

1. What plant did you use to make your dye? ______________________
2. Into which fabric sample was the dye best absorbed? ______________
3. Did the plant you chose produce the dye color you expected? ________
4. If you could repeat this investigation, what would you do differently? Why? ____

   ____

   ____

5. Would you expect your dye to be colorfast? ______________________

   Why or why not? ______________________________________

   ____

   ____

Owly's Appetite

Objective: Students will design and conduct a procedure for determining what an owl eats.

Time Required
Day 1: 55 minutes
Day 2: 55 minutes

Teaching Strategies In this partial inquiry, the feeding strategies of owls are examined. During this lab, use owl pellets. You can purchase them from a biological supply house or find them outdoors. If you find your own pellets, heat them in the oven at low temperatures for about one hour to kill bacteria.

Do not tell students what kinds of lab materials they should use but provide them with a variety of equipment including owl pellets, bowls of water, forceps, probes, and glue. Some other lab equipment that you might make available are small squares of cardboard, plain white paper, tape, scales, graduated cylinders, beakers, and funnels. Students may want to wear gloves, although they are not necessary.

Suggested Experimental Setup The easiest way to take apart an owl pellet is to soak it in a bowl of water. Each pellet will contain two to five bird and/or rodent skeletons. Let students work in groups to prevent this lab from taking more than two class periods.

As they remove a piece of the pellet from the bowl of water, students can pull out small bones and feathers, rinse them in clean water, and reassemble them on a sheet of paper. If they want to keep their skeletons intact, have them glue the skeletons in place.

Owl pellet kits are available from biological supply houses. In these kits, the pellets have been wrapped in aluminum foil and sterilized. They include diagrams of rodent skeletons to help students reassemble the owl's victims.

Background Information

Owly's Appetite

? ? ? ? ? Science Sleuth Question ? ? ? ? ?

What do owls eat?

Did you ever try to eat a bone or a hard seed? Most mammals and large birds cannot digest hair, bones, feathers, and hard seeds that they consume with their food. These materials are passed out of their bodies in their *scat* (feces). By inspecting an animal's scat, you can determine what that animal ate. This is not a very appealing job, but it does yield some useful information.

Owls and hawks have a unique way of handling their food wastes that cannot be digested. After a meal, they cough up the indigestible parts in a pellet that falls to the ground below their perch. Since owls usually feed from the same perch every night, you can sometimes find quite a few pellets under a perch.

Owls generally feed at night because they have no trouble seeing: their eyes are 100 times more sensitive to light than ours. In comparison to the rest of its head, an owl's eyes are very large. An owl must turn its entire head to see what is beside it because it cannot move its eyes around in their sockets.

Owls also have excellent hearing. They can hear extremely faint noises because of the heart shape of their faces. An owl's face is ringed with a ridge of feathers that serves to funnel sounds to the eardrums. Owl ears are long slits that are placed asymmetrically on the sides of the owl's head. This arrangement helps an owl determine the source of sound. Sound originating from below eye level will seem louder in the left ear. Sounds coming in above eye level seem louder in the right ear.

Figure 1. The heart-shaped face of the owl funnels sound to the eardrum.

Pre-Lab Questions

1. What kinds of materials would you expect to find in the scat of mammals and large birds? ______________________
   Why? ______________________

2. How do owls and hawks get rid of their indigestible waste? ______________

3. What might you determine about an owl or hawk by examining the pellets they regurgitate? ______________________
   ____

Name ______________________

Owly's Appetite

Objective: *Students will design and conduct a procedure for determining what an owl eats.*

Summary of Your Experimental Plan: ______________________

Materials Needed: ______________________

Procedure: ______________________

Data Table:

Remains Found in Owl Pellet	Description of Remains
#1	
#2	
#3	
#4	
#5	

Results:____

____

____

____

____

____

____

Conclusion Questions

1. Reassemble the skeletal remains of all animals found in your owl pellet. Glue these remains to a piece of notebook paper. How many animals are represented in this pellet? ________________________________

2. Most of the organisms that owls eat feed on plant parts or small animals. Draw a food web showing the owl, the small animals that the owl ate, and the plant parts that were probably food for the small animals.

3. A community is made up of a variety of living things in an area. How would the elimination of owls from the food web affect a forest or field community? __________

   ____

Bacterial Battle

Objective: Students will compare the number of bacterial colonies that can be grown from a clean dishcloth, a one-day-old dishcloth, and a two-day-old dishcloth.

Time Required
Day 1: 55 minutes
Day 2: 45 minutes
Day 3: 50 minutes

Teaching Strategies In this partial inquiry, students are provided with Background Information on bacteria and the conditions under which they grow. To make the lab realistic, let students wipe lunchroom tables or desk tops with dishcloths similar to those used in homes.

Students should select the lab equipment they need from the pieces that you make available to them. Items needed include sterile petri dishes containing agar, sterile swabs, clean dishcloths, toothpicks, paper plates, gloves, pens, and sterile forceps.

If students are not familiar with sterile technique, review it with them.

Suggested Experimental Setup Have students label the tops of the petri dishes as A, B, and C. Dampen a dishcloth and wipe it over table tops or desk tops. Open the stick-end of a sterile swab, remove the swab, and pass it over the dishcloth. Partially open petri dish A and gently rub the swab over the sterile agar. Reclose the petri dish and incubate. On Day 2, dampen the dishcloth, then repeat the swabbing procedure with a new sterile swab and petri dish B. On Day 3, repeat the entire activity using another sterile swab and petri dish C.

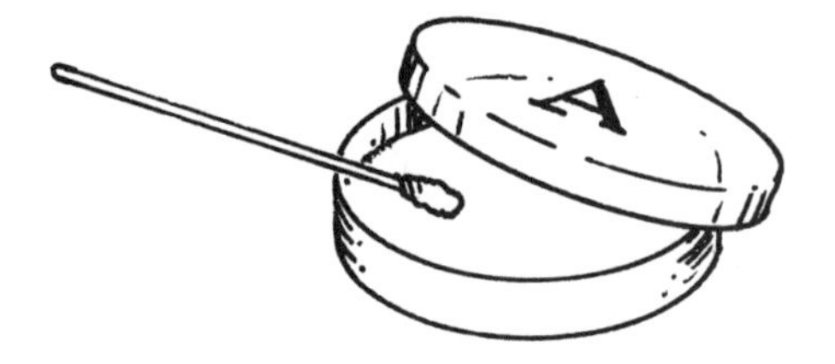

Figure 1. Label the petri dishes as A, B, and C. On Day 1, rub the sterile swab over a damp dishcloth and then over the agar in the petri dish.

Background Information

Bacterial Battle

? ? ? ? ? Science Sleuth Question ? ? ? ? ?

How quickly can bacteria grow?

Complete the following True/False survey to test your knowledge of bacteria.

____ 1. There are more bacteria than humans on earth.
____ 2. All bacteria cause disease.
____ 3. Several kinds of bacteria live in your body and help keep you healthy.
____ 4. Most bacteria can be killed by very high temperatures.
____ 5. Bacteria must have water to live.
____ 6. Bacteria grow well in warm, moist places.

Do you think of bacteria as "good" guys or "bad" guys? Actually, they are both. Bacteria can be found all over the earth—from the bottoms of hot springs to the tops of glaciers. There are more bacteria on the earth than any other kind of living thing. If you counted the number of bacteria on your body and compared that total to the number of cells in your body, the bacteria would win!

Cooks and medical personnel try to minimize the number of bacteria in their workplaces. When you are dealing with food or with people's health, it is always important to keep the environment germ-free to eliminate any disease-causing agents that might be lurking about. Bacteria can be killed with heat and disinfectants. Cool, dry conditions also inhibit bacterial growth. The reason food is kept in refrigerators and freezers is to prevent bacteria from growing on it and causing it to spoil.

In this lab we will try to determine whether or not a kitchen dishcloth is a good place for bacteria to grow. Generally, people clean kitchen countertops with dishcloths that can be reused several times. Let's find out how clean those dishcloths are.

Pre-Lab Questions

1. Where can you find bacteria? ____________________

2. How can bacteria be killed?____________________

3. Why do we refrigerate food? ____________________
1. Which dishcloth showed the most bacterial growth?____________________

Name ____________________

Bacterial Battle

Objective: *Students will compare the number of bacterial colonies that can be grown from a clean dishcloth, a one-day-old dishcloth, and a two-day-old dishcloth.*

Summary of Your Experimental Plan: In this lab, you will pass a sterile swab over a clean, damp dishcloth, and then rub that swab across a petri dish containing sterile agar. You will repeat that procedure on Day 2 with a clean swab and another petri dish, and then again on Day 3.

Petri dishes should be kept in a warm place (behind a refrigerator or in an incubator) for three days so that bacterial colonies can develop.

Samples Taken from Dishcloth	Number of Bacterial Colonies
Day 1	
Day 2	
Day 3	

Materials Needed: ____________________

____

____

Procedure: ______________________________

____

____

____

____

____

____

____

____

____

____

____

____

Results: ______________________________

____

____

____

____

____

____

Conclusion Questions

2. How can bacterial growth in the kitchen be reduced? ______________________________

____

3. A hospital keeps its operating room at very cool temperatures. How would these cool temperatures affect bacterial growth in the operating room? ______________________________

____

4. Why was it important to use three different sterile swabs in this lab? ______________________________

____

Unit 2

Physical Science

Water Bobbers

Objective: Students will determine whether or not the buoyant forces of fresh and salt water are the same.

Time Required
Day 1: 55 minutes

Teaching Strategies

In this full inquiry, students are provided with Background Information that helps them understand the concepts of buoyancy and density. They should read the entire lab before beginning their work.

Provide a variety of materials for students, including cups, beakers, graduated cylinders, scales, small objects that will float, aluminum foil, small weights, salt, water, and marking pens.

Suggested Experimental Setup

Prepare a salt water sample by stirring 10 grams of table salt in 100 ml of water. This solution is much saltier than the ocean and should give students fairly dramatic results. Actually, its salinity is close to that of Great Salt Lake.

Have students determine the densities of fresh (tap) water and salt water. Density can be determined by measuring the volume of a water sample and then determining its weight. To find its weight, place an empty paper cup or beaker on a scale. Record the the weight of the cup (or beaker). Then add the water sample. Find the new weight of the water plus the cup; then subtract the weight of the cup from the weight of the water and cup to find the weight of the water. Divide that weight (mass) by the water's volume.

Knowing the densities of the water samples, ask students which water sample they expect to display the largest buoyant force. Test student responses by floating an object in one water sample and marking the water level on its side. Then try the same object in the other water sample.

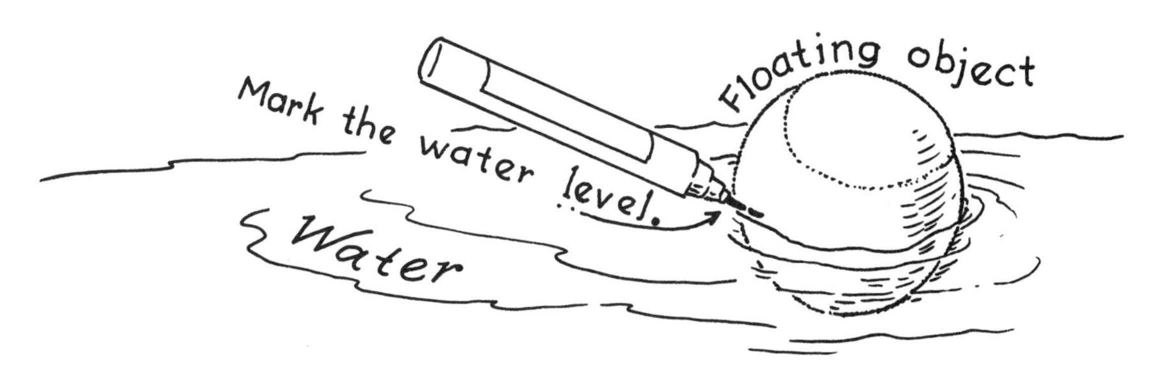

Figure 1. Mark the water level on a floating object.

Background Information

Water Bobbers

? ? ? ? ? **Science Sleuth Question** ? ? ? ? ?

Where is it easier for a swimmer to float: in fresh water or salt water?

Perhaps you have seen a television program that showed icebergs floating on the surface of the ocean. How do such heavy objects float? Two forces are acting on the iceberg. The first force, *gravity*, is directed downward. It is due to the weight of the iceberg on the water. The second and opposite force, *buoyancy*, pushes up and keeps the iceberg afloat.

Some objects float on water and some sink. When an object floats, there is more upward force acting on it than downward force. In other words, the force of gravity is less than the buoyant force. Objects sink when the weight of the object is greater than the buoyant force of the water on that object.

Thinking about a glass of ice water may help you understand these concepts. If you add ice to a glass of water, what happens to the level of the water? The water level rises as the ice pushes away or displaces some of the water. The amount of water displaced is equal to the buoyant force.

Figure 1. Two forces act on an object in water: gravity, a downward force, and buoyancy, an upward force.

To predict whether or not an object will sink or float in water, you need to know that object's mass and volume. By dividing the mass of an object by its volume, you can calculate its density. Water has a density of 1 gram/milliliter (g/ml). Objects with a density greater than 1 g/ml will sink in water, while those with a density less than 1 g/m will float on water. For example, a steel bolt with a density of 7.3 g/ml sinks in water. Balsa wood with a density of 0.12 g/ml floats on water.

Different liquids can also have variable densities. The more dense a liquid, the greater its buoyant force. How do the densities of ocean water and pond water compare? Do they have different buoyant forces? Perhaps you can devise an experiment to answer these questions.

Pre-Lab Questions

1. What force in water opposes the force of gravity on an object? ____________________

2. A small, hollow, ball-like object made of plastic can be used as a bobber on a fishing line. However, if that bobber were melted down into a small pellet of plastic, it would sink. In your opinion, why is this so? ____________________

____

____

Name ______________________

Water Bobbers

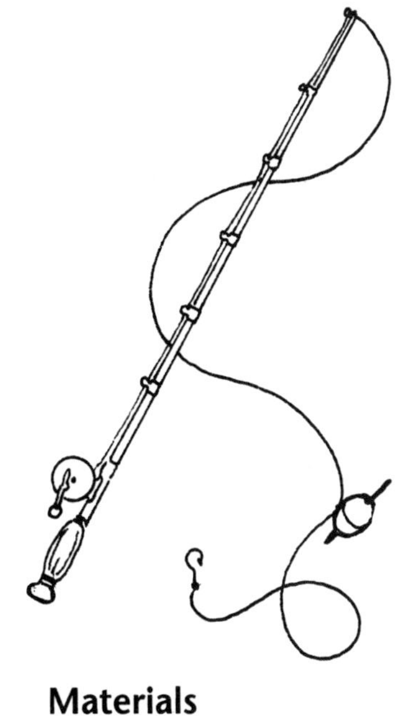

Objective: *Students will determine whether or not the buoyant forces of fresh and salt water are the same.*

Summary of Your Experimental Plan: ______________________

____

____

____

____

____

____

Materials Needed: ______________________

____

____

Procedure: ______________________

____

____

____

____

____

____

____

____

____

____

____

____

Data Table: (if needed)

Results: ______________________________

____

____

____

____

____

____

Conclusion Questions

1. Which is greater: the density of fresh water or the density of salt water? __________
2. Would it be easier to float objects on ocean water or in a pond? __________
3. When combining ocean water and fresh water, which will sink to the bottom of the container? __________
4. Which boat would float higher in the water if they were of equal volume and mass: the boat floating in the ocean or the boat floating on the pond? Explain your answer.

   ____

   ____
5. Define *buoyant force.* __________

   ____

   ____

   ____

   ____

Ink on the Run

Objective: Students will determine whether all water soluble black felt-tip pens yield the same color patterns when exposed to water.

Time Required
Day 1: 55 minutes

Teaching Strategies

In this partial inquiry, students are provided with Background Information that helps them develop a strategy for performing chromatography on ink.

Make several different water soluble, felt-tip pens of various brands available. Other materials that you might make available in the lab include pieces of filter paper or napkins, water, beakers, cups, tape, and scissors.

Suggested Experimental Setup

If you haves five pens, cut five pieces of filter paper into strips about three cm wide and ten to fifteen cm long (all strips should be the same size). Label the pens as A, B, C, D, and E. With a pencil, label the strips of paper in the same way (see Figure 1).

On the end of each strip opposite the label, use the appropriate pen to make a dot about two cm from the edge of the paper. Arrange the five strips in a cup or beaker containing a little water. (See Figure 1 in Background Information.) Do not let the ink dots touch the water. Watch them for 15 minutes.

Figure 1. Label each strip of paper. On the opposite end, place a dot of ink.

Students can measure the distance moved by each color; then they can measure the distance moved by the solvent (water) to calculate the F value of each color. To do so, follow this formula:

$$F = \frac{\text{distance moved by one color}}{\text{distance moved by water}}$$

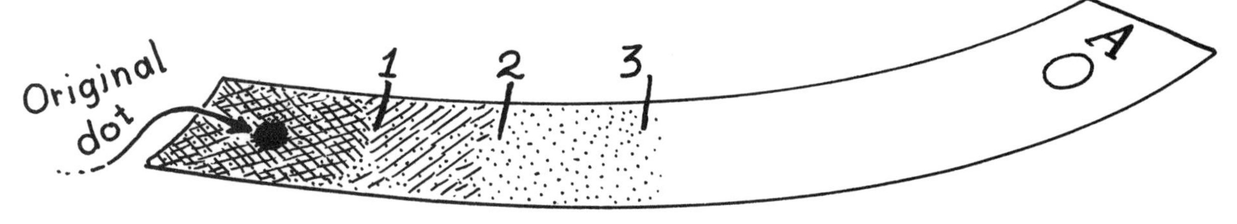

Figure 2. Measure the distance moved by each color. Divide that measurement by the distance moved by the solvent.

Background Information

Ink on the Run

? ? ? ? ? Science Sleuth Question ? ? ? ? ?

Can water change a note written in ink into a rainbow of colors?

A good friend wrote her new telephone number on a napkin with a felt-tip pen. During lunch, you accidentally spilled some water on the edge of a napkin. Thankfully, the water did not spill directly onto the ink and smear your message. However, a few minutes later, you notice that the water has traveled across the napkin and contacted the ink in the message. When it passed through the black ink, the water smeared it and changed it into several different colors.

The ink pen used to write the telephone number was water soluble. A substance that is water soluble can be dissolved in water. Some inks are water soluble. Black inks are usually combinations of several colors of dye, or coloring molecules.

When water contacted the ink on your friend's napkin, the coloring molecules dissolved in the water and were carried with the water across the napkin. The rainbow of colors was formed because different colored molecules have different shapes and sizes. The heavy molecules of some colors can only be carried a short distance. The lighter weight molecules of other colors travel farther. The result is a rainbow effect.

Because ink is a combination of two or more kinds of matter, it is considered to be a mixture. In a mixture, each type of matter keeps its own, unique characteristics. Mixtures can be separated by physical means, such as evaporation, filtration, and chromatography. If you want to separate ink into its component colors, you would choose the technique of chromatography. This procedure is widely used on dyes and drugs to find out what substances are in these mixtures.

Figure 1. A black ink spot is placed on filter paper. After contacting the bottom of the paper, water travels up it to merge with the ink. As the water travels through the ink and paper, it separates the ink into its component colors.

Pre-Lab Questions

1. Define *chromatography.* ______________________________

   ____

2. What causes a black felt-tip pen to yield bright colors when it is exposed to water?

   ____

3. How do you know that ink is a mixture? ______________________________

   ____

4. What would have happened to the message on the napkin if the ink had not been water soluble? ______________________________

   ____

Name ______________________

Ink on the Run

***Objective:** Students will determine whether all water soluble black felt-tip pens yield the same color patterns when exposed to water.*

Summary of Your Experimental Plan: ______________________

Materials Needed: ______________________

Procedure: ______________________

Data Table: (if needed)

Results: ____________________

____

____

____

____

____

____

Conclusion Questions

1. Name the different colors produced by the water soluble ink in your experiment.

____

2. Do all types of water soluble felt-tip pens produce the same pattern of colors? ______

3. Which brands of felt-tip pens produced the most colorful patterns? ______________

____

4. Describe how a police investigator could track down the manufacturer of a water soluble felt-tip pen that was used to write an illegal check. ______________

____

5. Describe how chromatography could be used to determine the types of colored pigments found in green plant leaves. ______________

____

What Is Your Reaction?

Objective: Students will determine whether heating a substance such as sugar or salt causes a chemical reaction to occur.

Time Required
Day 1: 55 minutes

Teaching Strategies In this full inquiry, students are provided with Background Information that helps them know how to recognize a chemical reaction when they see one. Based on that knowledge, students can devise a way to compare the effects of heating sugar and salt.

Students will need a hot plate or Bunsen burner, sugar, and salt. Make other lab items available to them, such as scales, graduated cylinders, aluminum foil, paper cups, and hot pads. Heating sugar in a beaker makes a mess that cannot easily be cleaned. Therefore, you may want students to heat their chemical on aluminum foil.

Suggested Experimental Setup Have students reiterate the evidences of chemical reactions so that they will know one when they see it. Then instruct them to heat a small sample of sugar on a piece of aluminum foil. After a few minutes, the sugar turns to black carbon, releasing carbon dioxide and water. Once the sugar has changed in appearance, students should stop heating it. CAUTION: Hot sugar can cause serious burns.

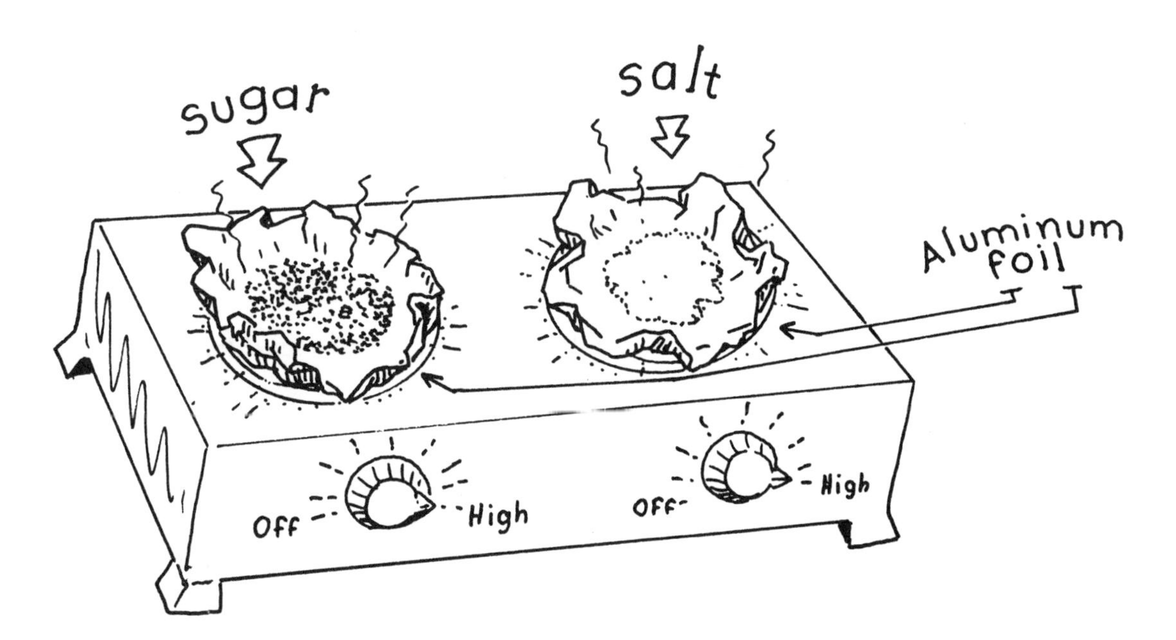

Figure 1. Heat samples of sugar and salt on aluminum foil. No matter how long they heat the salt, it will not change in any way.

Background Information

What Is Your Reaction?

??????? Science Sleuth Question ???????

Can heating a material cause it to undergo a chemical reaction?

What would happen if you were cooking and accidentally spilled vinegar into a container of baking soda? This blunder would result in a bubbling and hissing chemical reaction. The bubbling would indicate that a gas has been produced. In this case the gas is carbon dioxide.

All chemical reactions are not nearly as dramatic as the one described above, but they have one thing in common. In a chemical reaction, one or more new substances are created. The new substances have properties different from the original substances.

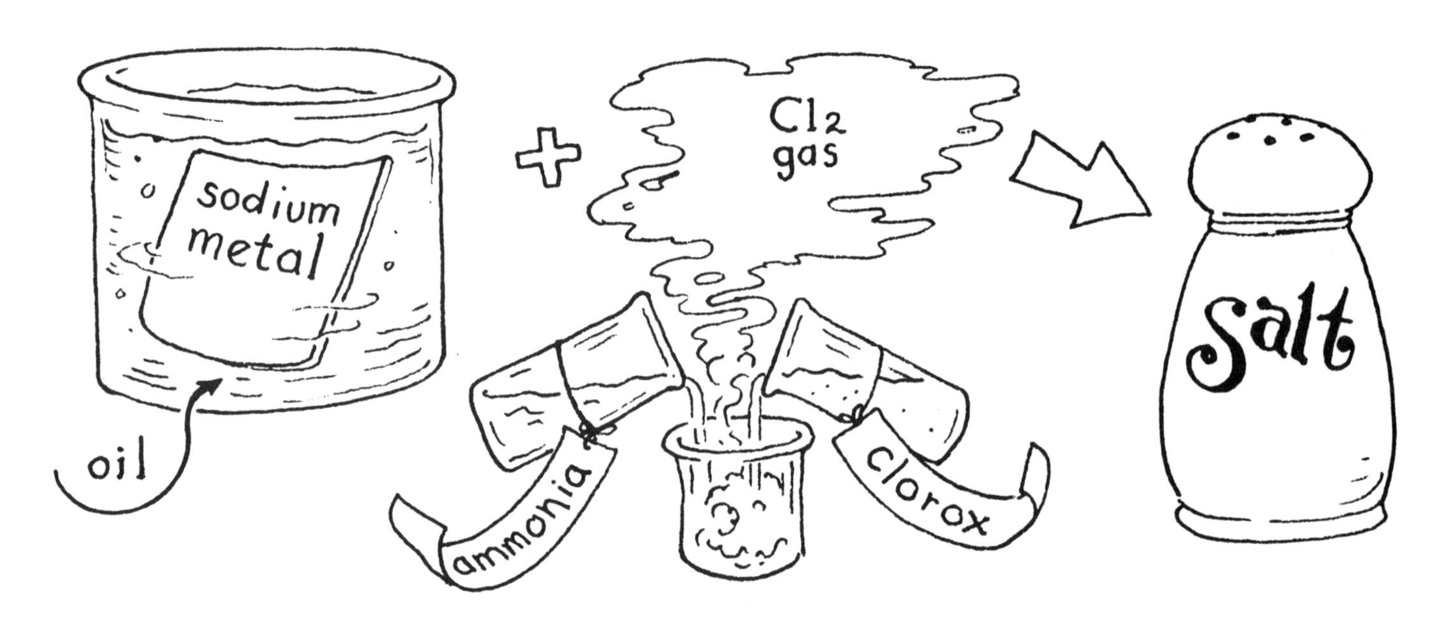

Figure 1. Sodium metal and chlorine gas can be combined to produce sodium chloride, or table salt. Sodium metal is very dangerous and causes severe burns. Chlorine gas is poisonous. Yet the product of the reaction of these two dangerous elements yields a safe, edible compound.

New substances form as a result of breaking old bonds and forming new ones. Of course, you cannot see chemical bonds breaking or being created, but there are some ways to tell when these changes occur. In the case of vinegar and baking soda, the bubbles indicated formation of a gas. This signaled that a chemical reaction had occurred.

Name ______________________

There are four clues that you can watch for as evidence of a chemical reaction. The first is the formation of a new solid when liquids are combined. This new solid is called a *precipitate*. For example, if you mix a clear liquid solution of silver nitrate with a clear liquid solution of sodium chloride, a white solid called silver chloride results. The second clue that a chemical reaction has occurred is color change, and the third is production of heat or light. Formation of a gas, as evidenced by bubbling, is the fourth clue.

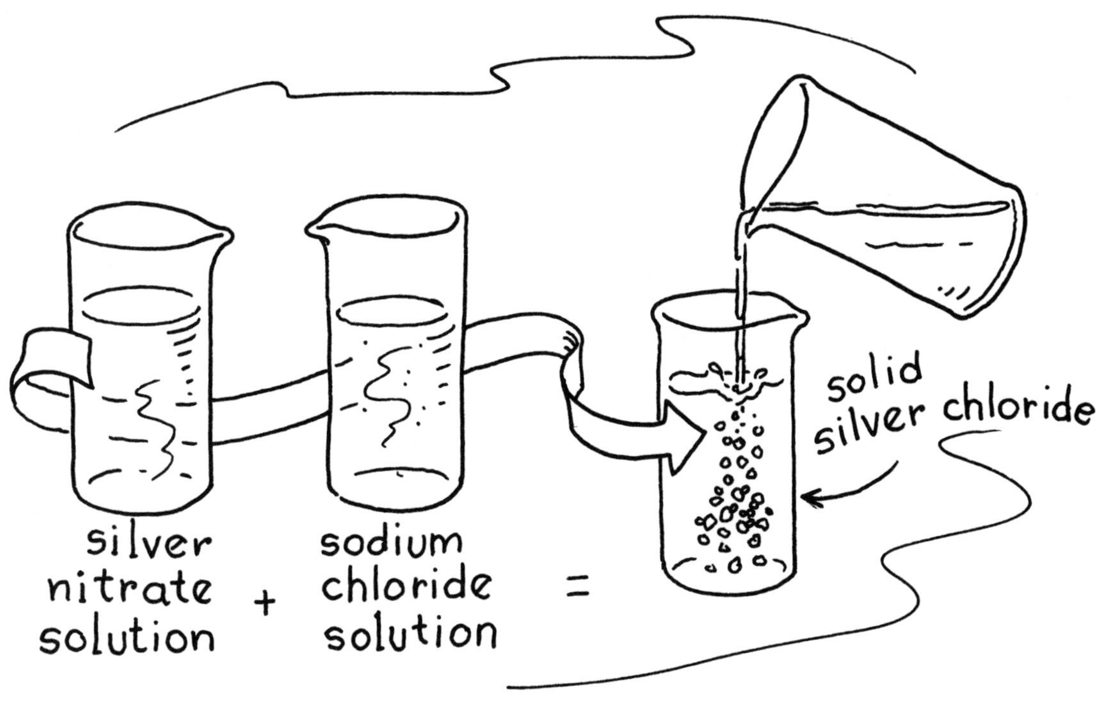

Figure 2. When you mix silver nitrate (as a clear solution) and sodium chloride (as a clear solution) together, a white solid called silver chloride results.

Pre-Lab Questions

1. Name the four indicators of a chemical reaction. ______________________

____

2. During a chemical reaction, what happens to the original substances? __________

____

3. During a chemical reaction, what happens to the chemical bonds of the original substances?

____

____

Name ______________________________

What Is Your Reaction?

Objective: *Students will determine whether heating a substance such as sugar or salt causes a chemical reaction to occur.*

Summary of Your Experimental Plan: ______________________________

Materials Needed: ______________________________

Procedure: ______________________________

Data Table: (if needed)

Results:____

Conclusion Questions

1. Table sugar is sucrose, whose chemical formula is $C_{12}H_{22}O_{11}$. The chemical formula for table salt or sodium chloride is NaCl. When these two chemicals were heated in the lab, which one(s) broke down into simpler substances? How do you know?

2. List the indicators of chemical reactions that you observed in this lab.

3. Of the two substances used in this lab, which one has stronger bonds holding its atoms or molecules together? Explain your answer.

4. If yeast, sugar, and warm water are mixed together in a soda bottle and a balloon is placed over the bottle, the balloon will inflate. Is this a chemical reaction? What is the evidence of this?

Plop, Plop, Fizz, Fizz

Objective: Students will determine how surface area affects the reaction rate of an effervescent antacid tablet with water.

Time Required
Day 1: 55 minutes

Teaching Strategies

In this full inquiry, students are provided with Background Information on factors that affect rates of chemical reactions. Students should read the entire lab before they begin so that they will anticipate the kind of information they are expected to gather. Emphasize that increasing surface area of a chemical increases reaction time. They can increase the surface area of the antacid tablet by crushing it (in a scrap of paper) or by cutting it into small pieces.

For student use in the lab, provide a variety of equipment including cups, beakers, water, effervescent antacid tablets, clock or stop watch, and a hammer or rock for crushing tablets.

Suggested Experimental Setup

Fill two small beakers or cups half full of water. With a stopwatch ready to begin timing, drop a whole antacid tablet in the first cup. Time the bubbling. With a pen or piece of tape, mark the highest point reached by bubbles on the cup.

Chop or crush the second antacid tablet. With a stopwatch ready to begin timing, drop the chopped tablet in the second cup. Time the bubbling and again mark the highest point reached by bubbles on the cup.

To extend the experiment, have students predict the behavior of an antacid tablet that is cut into four pieces. Then have them test that tablet in a half cup of water to see whether their predictions were correct.

Figure 1. Place a whole effervescent antacid tablet in one cup. Time the bubbling and mark the highest point reached by bubbles. Repeat the procedure with a crushed tablet.

Background Information

Plop, Plop, Fizz, Fizz

? ? ? ? ? **Science Sleuth Question** ? ? ? ? ?

How does surface area affect the rate of a chemical reaction?

The world is constantly changing. Changes can be classified into two types: physical and chemical. Physical changes are those that rearrange matter but do not create a new kind of matter. In a chemical change, a new kind of matter is created.

Some physical changes are pretty unspectacular. For example, if you tear a piece of paper into six strips, the paper undergoes a physical change. You have paper when you begin, and it is still paper after you rip it. So, you did not create anything new by tearing the paper.

However, you can chemically transform the paper into something else. If you burn the paper, it is not paper anymore. Instead, the paper is changed to water vapor, carbon dioxide gas, and ash. In the terms of chemistry, the original paper was the reactant, and the water vapor, carbon dioxide, and ash are the products.

Figure 1. Cutting paper is a physical change. Burning paper is a chemical change.

In this lab, we will examine the way surface area affects the rate of a chemical reaction. To do so, we will observe the reaction between an effervescent antacid tablet and water. This reaction is a simple one to monitor because bubbling indicates that a reaction is occurring. Once bubbling has stopped, the chemical reaction has stopped.

Surface area refers to the amount of a reactant that is available to interact with other reactants. For example, let's say that your task is to color an entire potato red. You peel a potato and observe the soft white outer surface, which is the potato's surface area. If you drop that whole potato into a bottle of red dye, the dye can only interact with the surface area of the potato. It may be several days before any of the dye soaks all the way to the center of the potato.

A faster way to dye all of the potato red is to increase the amount of its surface area that is exposed to the dye. You can do this by cutting the potato into small pieces before placing them in the dye. That way, the red dye will only have to travel a short distance in each piece. You could dye this potato faster than the whole one.

Figure 2. Cutting a potato into small pieces increases its surface area.

Pre-Lab Questions

1. How could a raw egg be physically changed?______________________
   Name one way that it could be chemically changed. ______________________
   ____

2. Why do cooks chop up onions before adding them to vegetable soup? __________
   ____

3. Explain the difference in a chemical change and a physical change. __________
   ____
   ____

4. You have a ball of clay that you are going to drop into a vat of acid. What is the surface area of that ball of clay? How could you increase the surface area of the clay?
   ____
   ____

5. Place an "x" in the blank beside the reaction that will occur faster:
   ______ a. Soaking a whole leaf in alcohol to extract the pigment
   ______ b. Soaking a crushed leaf in alcohol to extract the pigment

   ______ c. Swallowing and digesting a whole aspirin tablet
   ______ d. Swallowing and digesting a chopped aspirin tablet

   ______ e. Burning coal dust in a furnace
   ______ f. Burning small pieces of coal in a furnace

Name ____________________

Plop, Plop, Fizz, Fizz

Objective: *Students will determine how surface area affects the reaction rate of an effervescent antacid tablet with water.*

Summary of Your Experimental Plan: ____________________

Materials Needed: ____________________

Procedure: ____________________

Data Table: (if needed)

Results: ______________________________

____

____

____

____

____

____

Conclusion Questions

1. What is the chemical reaction you are investigating in this lab? ______________________________

   ____

2. What are the reactants in the chemical reaction you are investigating in this lab?

   ____

3. How can you determine whether or not this chemical reaction is occurring? ______________________________

   ____

4. Which is easier to set on fire: a log or a pile of wooden splinters? Why? ______________________________

   ____

   ____

5. Which piece of candy will dissolve in your mouth faster: the whole piece or the crushed piece? Why? ______________________________

   ____

   ____

Putting on the Heat

Objective: Students will determine whether temperature affects the rate of a chemical reaction.

Time Required
Day 1: 55 minutes

Teaching Strategies In this full inquiry, students are provided with Background Information that helps them explore ways that the rates of chemical reactions can be altered.

Students should be provided with effervescent antacid tablets, beakers, hot plates, ice, water, thermometers, and stopwatches.

Suggested Experimental Setup Students can create cold water by adding ice to tap water. Hot water may be available at the school, or you might have to heat water for students.

Label two beakers, one as A and the other as B. Place cold water in beaker A. With a stopwatch, determine how long it takes the effervescent tablet to stop reacting with the water. Record this time.

Place hot water in beaker B. Add the effervescent antacid tablet and time the reaction. Compare the two times, and determine in which case the reaction occurred faster.

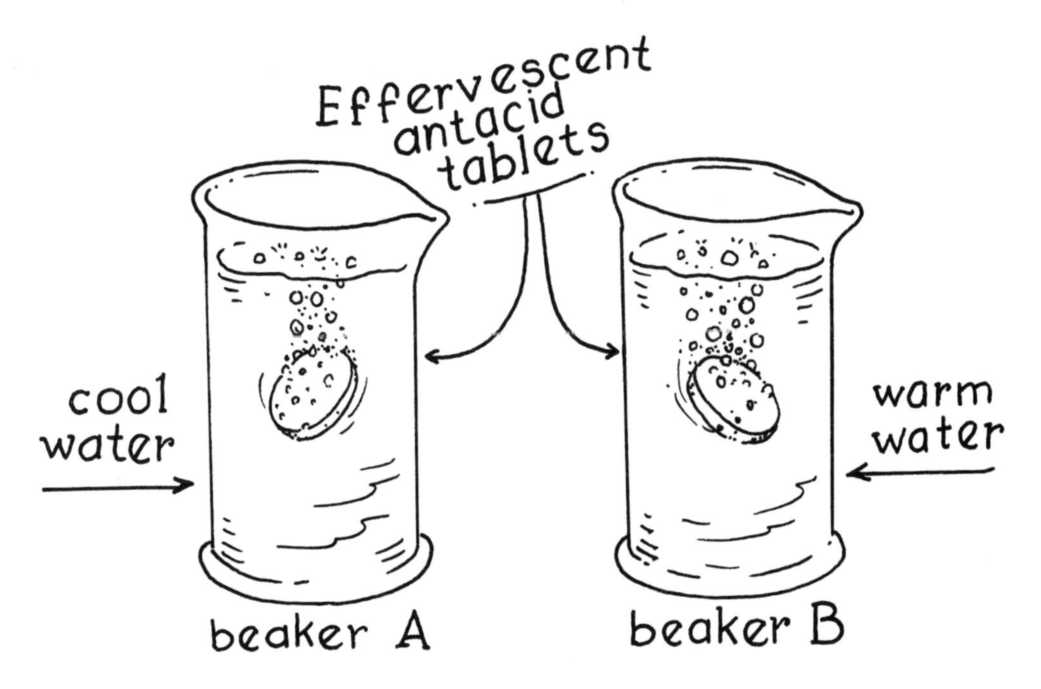

Figure 1. Drop an effervescent antacid tablet in beaker A (cool water) and one in beaker B (warm water). Time the reaction of the tablets and the water.

Background Information

Putting on the Heat

? ? ? ? ? Science Sleuth Question ? ? ? ? ?

How does temperature affect the rate of a chemical reaction?

A chemical reaction occurs when molecules collide with enough energy to break old chemical bonds and form new ones. Some reactions, like explosions, occur quickly. Other reactions, such as the rusting of metal, occur slowly. It is possible to change the speed at which some reactions occur.

After using butter on your baked potato, you probably place the butter in the refrigerator to keep it from spoiling. In effect, you use a cold environment to slow the rate of the reaction that might sour the butter. In industry the reverse is often true: scientists want to speed the rates of chemical reactions. By doing so, they can reduce manufacturing costs.

There are many factors that can affect the speed at which a reaction takes place:

a) Change in pressure—When the pressure of a gas is increased, the individual particles of the gas are squeezed together. As they move closer together, they are more likely to contact each other and react.

b) Change in temperature—When the temperature of a solution increases, the individual particles collide more often and with more energy, increasing the speed of their reaction.

c) Change in concentration—The concentration of a solution can be important in determining reaction rate. A match will ignite faster in an oxygen-rich environment than it will in an oxygen-poor one.

d) Change in surface area—The greater the surface area of a reactant the faster the chemical reaction. Crushing, chopping, or mashing a reactant can speed reaction rate.

e) Change in exposure to light—Exposure to light can cause reaction rates to increase. Some chemicals, such as hydrogen peroxide, decompose when exposed to light. For this reason they are stored in a dark bottle to block out light.

Pre-Lab Questions

1. Name some factors that can alter the speed of a chemical reaction. ________________

____

2. In order for a reaction to occur, what must happen between molecules? ____________

____

3. Explain why the thin strips of potato that we call french fries cook faster than the entire potato. ____

____

Name ____________________

Putting on the Heat

Objective: *Students will determine whether temperature affects the rate of a chemical reaction.*

Summary of Your Experimental Plan: ____________________

Materials Needed: ____________________

Procedure: ____________________

Data Table: (if needed)

Results: ______________________________

Conclusion Questions

1. How did you determine the speed with which the reaction occurred? ______________________________

2. How does increasing the temperature affect the reaction of an effervescent antacid tablet in water? ______________________________

3. Based on the results of this lab, explain why animals that died and were frozen in the snow did not decay. ______________________________

4. For yeast to undergo respiration, it must be mixed with a carbohydrate and water. Do you think this process works better in cool or warm water? ______________________________
   Explain your answer. ______________________________

Unit 3

Earth Science

Ice Sculpture

Objective: Students will assess the effects of ice moving over a mound of soil on a slope.

Time Required
Day 1: 55 minutes
Day 2: 30 minutes

Teaching Strategies In this partial inquiry, students are provided with Background Information that guides them as they develop an idea for observing glacial effects on soil. Have students read the entire lab before they begin so that they will anticipate the kind of information they are expected to gather.

Students can use a variety of materials in this lab. Provide them with soil samples (sand, loam, clay, etc.) and something to place them in, such as metal pie plates, cookie sheets, or strips of cardboard. (You can extend this lab by comparing glacial effects on different kinds of soil. For this lab, sand may be the best soil to use.) Other lab materials can include ice cubes, chips of ice, rulers, blocks of wood, and an assortment of equipment unrelated to the lab.

Suggested Experimental Setup Cut a piece of cardboard that is about 24 inches long and 8 to 10 inches wide. Place 100 ml (by volume) of sand on one end of the cardboard. Elevate this same end by propping it on a book or a block of wood.

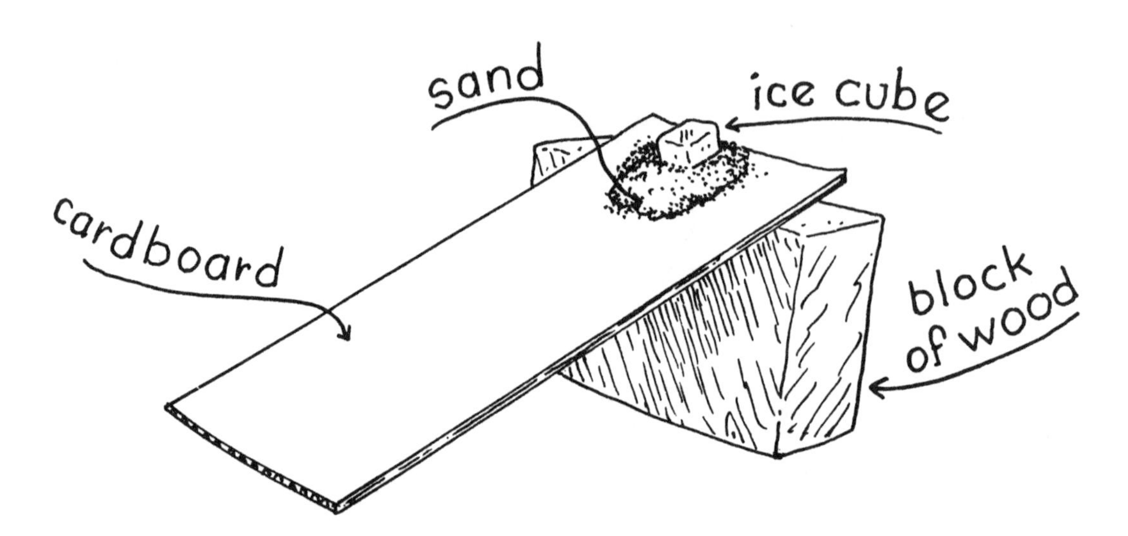

Figure 1. Place 100 ml of sand on one end of a piece of cardboard. Elevate that end of cardboard by placing a book or block of wood under it.

Place an ice cube on the soil and observe how it affects the soil for 20 minutes, recording information about changes in soil position every 10 minutes. Leave the soil-ice-cardboard setup in place until the next day. Again, record any changes in the position of the soil.

Background Information

Ice Sculpture

? ? ? ? ? **Science Sleuth Question** ? ? ? ? ?

Can a glacier change the appearance of the land?

Most of us have never seen a glacier. Glaciers are only found in the coldest regions of the earth, such as the north and south poles, and very high up in the mountains. Glaciers are gigantic masses of ice that flow slowly across the land. They can be several miles long and 300 to 10,000 feet thick.

Glaciers begin to form when more snow falls in the winter than melts in the summer. Over time, this excess snow builds up into thick layers. The snow crystals on the very bottom of these layers are crushed together by the weight above them. Eventually, the snow flakes and ice crystals are compressed into a solid sheet of glacial ice. The ice becomes so thick that it moves under the pressure of its own weight.

Gravity affects glaciers as it does everything else and causes them to flow down a slope. The movement of a glacier along its path is extremely slow, usually less than a foot each day. Even though the surface of a glacier is rigid, ice crystals on the underside shift and slide over one another as the frozen mass moves. The friction of this movement warms and melts some of the ice along the base, sending frigid water into the soil where it refreezes.

Figure 1. The parts of a glacier

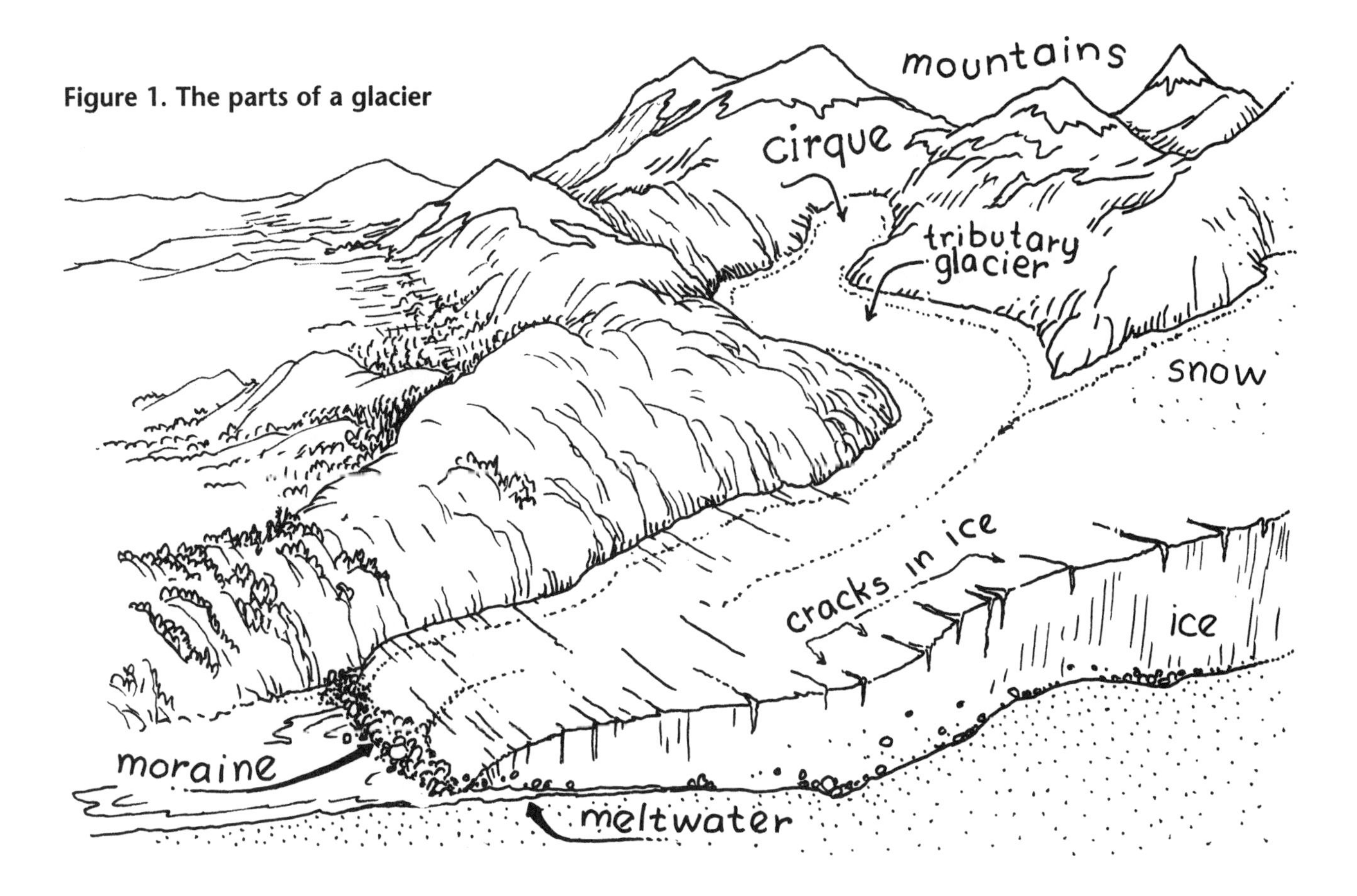

Name ______________________________

As glaciers travel over the land, they affect the land's shape and appearance in two ways:

a. by eroding the soil, and

b. by transporting rock and debris.

Glacial erosion occurs when the moving ice mass, dragging rocks along its base, grinds the bedrock down into a smooth surface. As glaciers melt and recede, the rocks and debris that traveled along with them are left in piles or ridges called *moraines.* Often, the rocks in moraines are rounded and smooth because of the tumbling they experienced in their travels.

Pre-Lab Questions

1. What is a glacier? ______________________________

2. How do glaciers affect the land? ______________________________

3. Why are the rocks in moraine rounded? ______________________________

____

Name ____________________

Ice Sculpture

Objective: *Students will assess the effects of ice moving over a mound of soil on a slope.*

Summary of Your Experimental Plan: ____________________

____

____

____

Data Table 1: Predictions of How Ice Will Affect a Mound of Soil

Time	Appearance of soil before, during, and after ice moves over it
Before soil is exposed to ice	
10 minutes after soil is exposed to ice	
20 minutes after soil is exposed to ice	
24 hours after soil is exposed to ice	

Materials Needed: ____________________

____

____

____

____

Procedure: ____

____

____

____

____

____

____

____

____

____

____

____

Data Table 2: Actual Results from Experiment

Time	Appearance of soil before, during, and after ice moves over it
Before soil is exposed to ice	
10 minutes after soil is exposed to ice	
20 minutes after soil is exposed to ice	
24 hours after soil is exposed to ice	

Results: ____

____

____

____

Conclusion Questions

1. How did the ice affect the height of the mound of soil? ______________________

____

2. How did your predicted results compare with the actual results? ______________

____

3. How did the ice affect the width and/or length of the mound of soil? __________

____

4. Did the ice in your experiment create any moraines? ______________________
   If so, how tall were they? ____

5. Do you live in an area where a glacier has been? _________________________
   How do you know? ____

____

Soil on the Move

Objective: Students will examine and quantify the differences in soil erosion by water on a gentle slope and on a steep slope.

Time Required
Day 1: 55 minutes

Teaching Strategies In this partial inquiry, students are provided with Background Information to help them develop a plan for determining how slope affects rate of soil erosion by water. They should read the entire lab before beginning so that they will anticipate the kind of information they are expected to gather.

Students should select the lab equipment they need from the pieces that you make available to them. Provide them with sand for their soil samples, cardboard for building slopes, paper cups or beakers to hold water, graduated cylinders or beakers for measuring volume, and an assortment of unrelated material.

Suggested Experimental Setup Remind students that when they quantify an experiment they must limit their variables. Make certain that members of each lab group measure their original volume of soil at the top of the slope. By then measuring the amount of soil that is moved by water, they can compare the effects of slow-moving water and of fast-moving water.

Measure 50 ml (volume) of soil and set it on the high end of a cardboard slope. Label this piece of cardboard as A. Create another cardboard slope with 50 ml of soil and label it as B. Make certain that slope B is steeper than slope A.

Figure 1. Slope B is steeper than slope A.

In the side of a paper cup, use the lead of your pencil to make a small hole. Place your finger over the hole; then pour 100 ml of water into the cup. Hold the cup of water directly over the soil in setup A; then remove your finger. Repeat the same procedure with the soil in setup B.

Measure the farthest distance that soil was carried down each slope. Measure the amount of soil that moved down slope A and slope B.

Background Information

Soil on the Move

? ? ? ? ? Science Sleuth Question ? ? ? ? ?

Why does soil erosion occur faster in some places than others?

Soil rarely stays in one place. It can be moved by wind, water, and ice in a process called *erosion*. Even though erosion is a natural process, activities by people have unnaturally increased its rate.

Any time we disturb the soil by digging, plowing, or bulldozing, we increase the rate of erosion. When the ground is covered with grasses, trees, shrubs, vines, or any other types of plants, erosion rarely occurs. The plants with their roots and associated dead plant matter on the ground help hold the soil in place.

As water flows down a slope that is covered with loose soil, it picks up much of the soil and carries it to the bottom of the hill. That is why farmers in hilly locations take special measures to curb the flow of water down hills. Slowing the water gives it enough time to soak into the ground. One way farmers protect hillsides is by plowing rows around the hill instead of up and down the slope. Another anti-erosion technique is to keep the hill covered in vegetation.

Figure 1. By plowing rows around a hill, instead of down it, farmers can prevent soil erosion.

The speed of wind and water across soil affects the rate of soil erosion. A gentle breeze blowing across a freshly plowed field will rarely move much of the topsoil. However, a strong wind can lift hundreds of pounds of soil and move it to a new location. In the 1930s farmers in the midwestern United States lost the topsoil of their rich farmland. After plowing the soil and preparing it for planting, strong winds blew across the area and removed all of the valuable soil. Farmers were left with barren soil that was useless for farming.

Pre-Lab Questions

1. With an "x," indicate the activities that expose soil and increase the rate of erosion:

____ a. building a house
____ b. plowing furrows to plant a garden
____ c. constructing a new driveway
____ d. tilling the yard to plant grass
____ e. building a highway
____ f. mowing grass
____ g. clear-cutting a forest
____ h. constructing a bridge

2. What is erosion? ____________________

____

3. How do farmers in hilly locations slow soil erosion? ____________________

____

Name ______________________

Soil on the Move

Objective: *Students will examine and quantify the differences in soil erosion by water on a gentle slope and on a steep slope.*

Summary of Your Experimental Plan: ______________________

____

____

____

____

Data Table:

Soil	Volume of soil in milliliters
In original location	
Moved by water flowing slowly	
In original location	
Moved by water flowing rapidly	

Materials Needed: ______________________

____

____

____

____

Procedure: ______________________________

____

____

____

____

____

____

____

____

____

Results: ______________________________

____

____

____

____

____

____

Conclusion Questions

1. What happens to the speed of the water as the slope is increased? ______________
2. As the speed of flowing water increases, do you think the water's energy increases or decreases? ______________________________
   Why? ______________________________
3. How is it possible for flowing water to move dirt? ______________________________
   ____
4. It is important in any experiment to have as few variables as possible. The change in slope is the variable that we are testing in this experiment. What are some other variables that you could test in an experiment like this? ______________
   ____
5. Explain why you agree or disagree with the following statement: An experiment should test only one variable at a time. ______________
   ____
   What variable are we testing in this lab? ______________

The Frozen Explosion

Objective: Students will determine whether or not ice is strong enough to break a rock by measuring how much water expands when it freezes.

Time Required
Day 1: 55 minutes

Teaching Strategies In this partial inquiry, students are provided with Background Information that helps them understand the behavior of water as it freezes. They should read the entire lab before they begin so they will anticipate the kind of information they are expected to gather.

For student use in the lab, provide a variety of equipment including water, graduated cylinders, paper cups, markers, rulers, thermometers, and a freezer or ice chest.

Suggested Experimental Setup Place some water in a small graduated cylinder (10-20 milliliters). On the Data Table record the amount of water in the graduated cylinder and its temperature. Place the cylinder and water in a freezer or ice chest. Check the water's volume and temperature every five minutes.

If you do not have enough graduated cylinders for the class, let students make their own graduated paper cups. Pour 15 ml of water in the cup; then mark and label the water level. Pour another 5 ml of water into the cup and mark the level again. Students should mark the levels for 15, 20, 25, and 30 ml.

Background Information

The Frozen Explosion

? ? ? ? ? **Science Sleuth Question** ? ? ? ? ?

Is ice strong enough to break a rock?

Have you ever wondered where all of our dirt came from? Has it always been here? Probably not. Dirt is an accumulation of a lot of things, such as dead organisms and eroded pieces of rock. Early earth was all rock, but that rock has been worn down into smaller pieces by the action of wind, water, ice, and other factors.

Water can exist in all three phases of matter—solid, liquid, and gas. In each of these phases, water has a lot of strength. You have heard about the devastating effects of liquid water when floods sweep across land and carry away everything in their paths. Liquid water's strength can also be beneficial. It can be used to perform work, such as turning wheels to generate electricity.

The same is true of water in the gaseous state. Whether we call it steam or water vapor, this gaseous compound accomplishes a lot of work in places like power plants and steam engines. But did you know that water in the solid state can also change things? The strength of frozen water is due to the fact that water behaves strangely when it freezes.

Most liquids contract when they cool. This makes sense when you remember that everything is made up of atoms and molecules that are in constant motion. The warmer the environment of the molecules, the faster those tiny particles move. Cooling causes atoms and molecules to slow down. And once they slow down, they can get closer together.

When water cools, its molecules slow down, too. However, water's molecules do not get very close together. Because water is a polar compound, the molecules arrange themselves in a pattern that takes up more space than the molecules required in the liquid state.

Polar molecules can be compared to bar magnets. Bar magnets have a positive end and a negative end. If you tried to pack several bar magnets in a box, you would have to arrange them in that box so that the negative end of one magnet is near the positive end of another. You could not place two positive ends or two negative ends together. That is because like charges on magnets repel each other, but opposite charges attract. It would be impossible to pack bar magnets with like poles together.

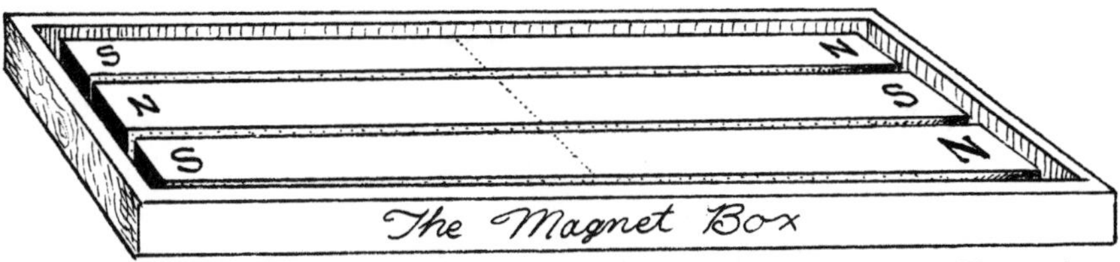

Figure 1. Bar magnets can be packed in a box with unlike poles together. Unlike poles attract each other, but like poles repel.

A water molecule behaves very much like a bar magnet. When the molecules are moving slowly and can get close to one another, the negative end of one water molecule is attracted to the positive end of another. As a result, the water molecules arrange themselves in a lattice-like structure that actually takes up more space than those same molecules required in the liquid state.

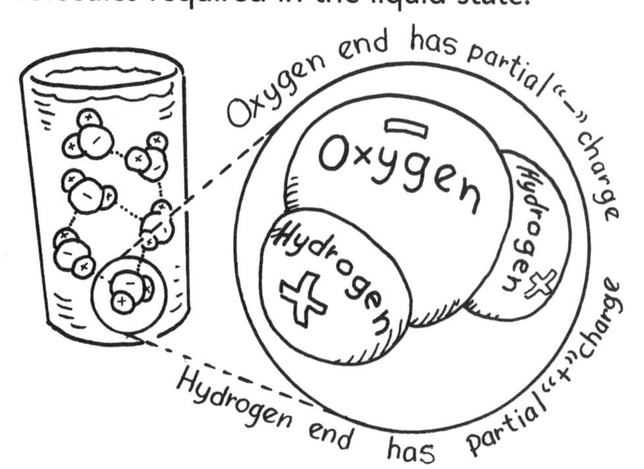

Figure 2. Water is a polar molecule. It has a positive end and a negative end.

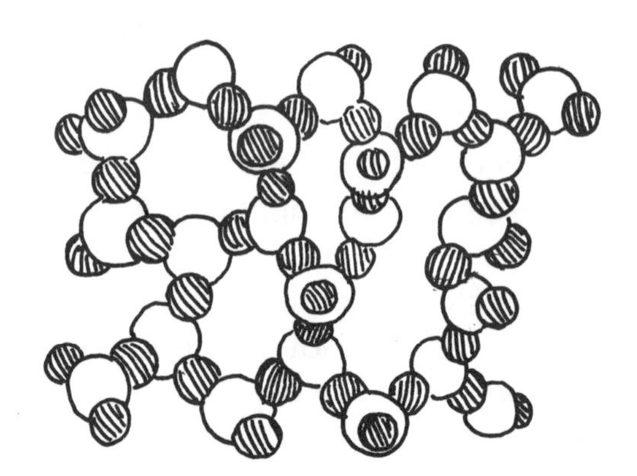

Figure 3. When water freezes, the molecules form a lattice-like arrangement so that the negative end of one water molecule is near the positive end of another water molecule.

As water freezes, it exerts quite a bit of pressure to get into that lattice-like arrangement. Have you ever placed a bottle full of water or soft drink in the freezer? That makes a mess! As the water freezes and the molecules get into their lattice arrangement, they position themselves so that a positive part of one molecule is near the negative part of a nearby molecule. The resulting frozen water takes up more space than it did as a liquid.

The strength of frozen water can be seen in the winter. Water in the liquid state can soak into concrete roadways or rocks. When that liquid freezes, it expands. Imagine how the expansion of water affects the roadways and rocks.

Pre-Lab Questions

1. What are the three states in which water can exist? ______________________

2. In which state are water molecules moving fastest:
   a. solid or liquid?
   b. solid or gas?
   c. liquid or gas?

3. How does freezing water contribute to the erosion of rocks? ______________________

   ____

4. What are polar molecules? ______________________

   Give an example of a polar molecule. ______________________

   ____

Name ____________________

The Frozen Explosion

Objective: *Students will determine whether or not ice is strong enough to break a rock by measuring how much water expands when it freezes.*

Summary of Your Experimental Plan: ____________________

Materials Needed: ____________________

Procedure: ____________________

Data Table: Volume of Water as It Freezes

Temperature	Volume of Water

Results: ____

Conclusion Questions

1. If you place a canned soft drink in the freezer, what will happen? __________
   Why? ____________________________________
2. Rocks contain pores and crevices that can hold water. What happens to these rocks when water in those pores freezes? _________________________
3. In what part of the world would you expect to find the most erosion due to cold weather: in the tundra, which is dry and cold all year round, or high in the Rocky Mountains, where the weather is warm in summer and cold in winter? Why?
4. Explain why water expands when it freezes. ____________________

5. If you are exposed to cold weather for a long time without any gloves to keep your hands warm, the water in the cells of your hand can freeze. How do you think that this freezing will affect the cells? ______________________________

____

6. In the following graph, show how the volume of water changes as the temperature of the water decreases.

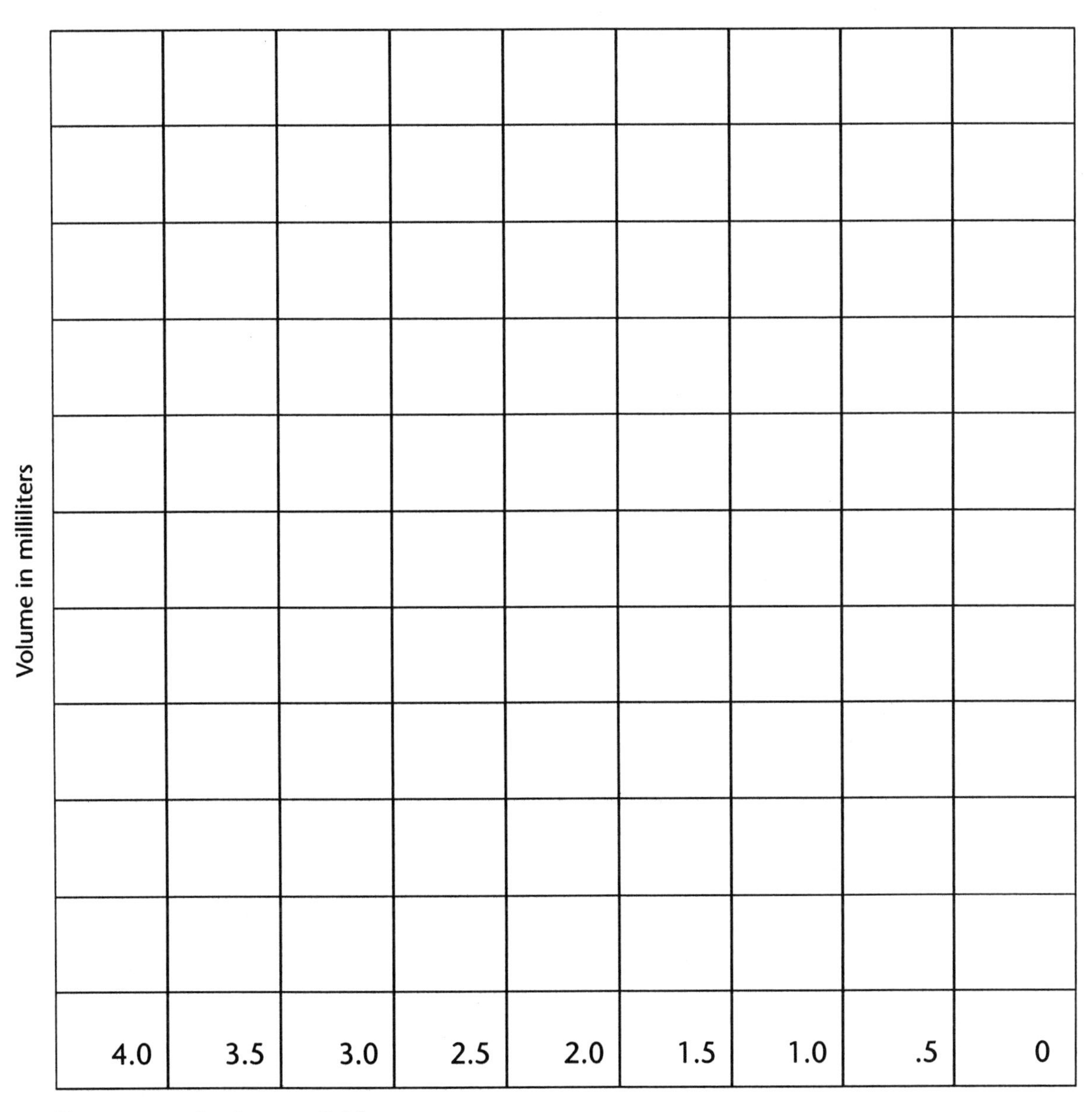

Don't Drink the Water

Objective: Students will clean a sample of dirty water.

Time Required
Day 1: 55 minutes

Teaching Strategies In this partial inquiry, students are provided with Background Information on steps used to purify drinking water. In their own investigation, students will not need to use all of these steps. Inform the class that an efficient municipal water system cleans its water in as few steps as possible, primarily to save money. You could extend this lab by charging students some amount of play money for the equipment they use. The class can then discuss who got their water the cleanest at the least expense.

For student use in the lab, provide a variety of materials and equipment including paper or plastic cups, droppers, pipettes, funnels, beakers, water, charcoal, paper towels, sand, gravel, and alum. If you want students to distill their water, also provide distillation tubes, Erlenmeyer flasks, stoppers, and burners.

You can give students' purified water the ultimate cleanliness test with a conductivity meter. Even the clearest-looking water will still contain some minerals and, therefore, will conduct a current.

To make "polluted" water, blend or briskly stir together water, a few drops of oil, one clove of garlic, and a little ground coffee. Give each group of students a small paper cup of this water.

Suggested Experimental Setup Add a pinch of alum to the sample of polluted water, stir gently, and set the water on a desk or table for a few minutes. Decant or pipette the oil from the top of the water and discard. Decant the water from the sediment. Filter the water through gravel and then through sand. A filter can be made by punching a few small holes in the bottom of a paper cup and then adding gravel to the cup. A similar filter can be made with sand in a cup. Finally, filter the water over aquarium charcoal to remove color and odor.

If you want students to distill their clean water samples, use the following setup:

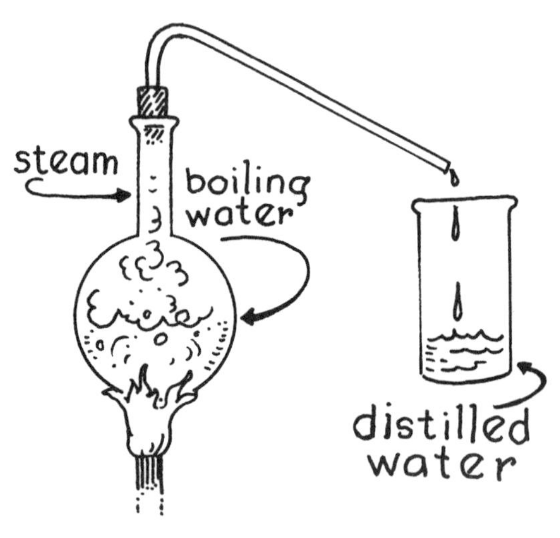

Figure 1. a distillation apparatus

Background Information

Don't Drink the Water

? ? ? ? ? Science Sleuth Question ? ? ? ? ?

Can you clean a sample of dirty water?

How much water do you drink every day? Many nutrition experts tell us that we should drink eight 12-ounce glasses daily. After all, our bodies are about 60% water. We have to keep replacing what we use.

Clean drinking water is something that most of us take for granted. When there were only a few people on the earth, water in streams and springs was safe to drink. However, in these crowded times, most waterways are polluted with litter, gasoline, acid rain, and raw sewage.

You probably get your clean drinking water from a city or county water treatment plant. This type of facility takes ground or surface water, removes the trash, minerals, sediment, and foul odors, and then kills the bacteria. Depending on the original quality of the water, one or more of the following methods may be used:

a. Settling—By holding the water still for a few hours, soil and other solid matter can settle out. Settling can be aided by the addition of alum, a chemical that pulls clay and other particles into clumps and speeds the process.
b. Filtration—When water is passed through a porous material, anything larger than the pores is filtered out.
c. Adsorption—Some materials, like activated charcoal, can adsorb and remove foul odors from water.
d. Distillation—Water is evaporated and then condensed in another container. Only the water itself can evaporate, so all pollutants are left behind.
e. Disinfection—Water is treated with chemicals that kill disease-causing germs.

Name ____________________

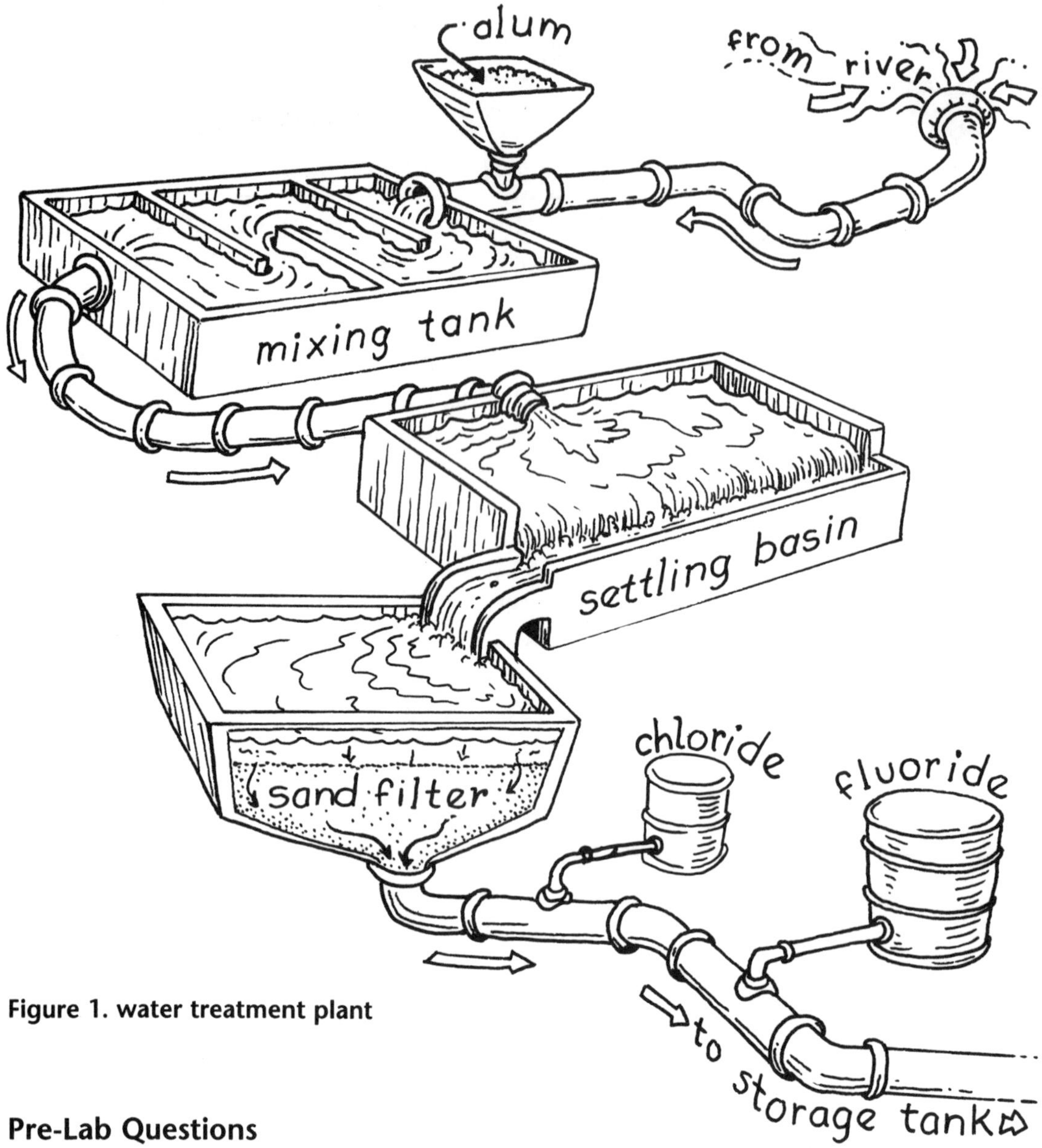

Figure 1. water treatment plant

Pre-Lab Questions

1. Why must our drinking water be treated before we can use it? ____________________
____

2. List five steps that can be part of a water purification system. ____________________
____
____

3. Why is alum added to water processed in the water treatment plant? ____________________
____

Name ____________________

Don't Drink the Water

Objective: *Students will clean a sample of dirty water.*

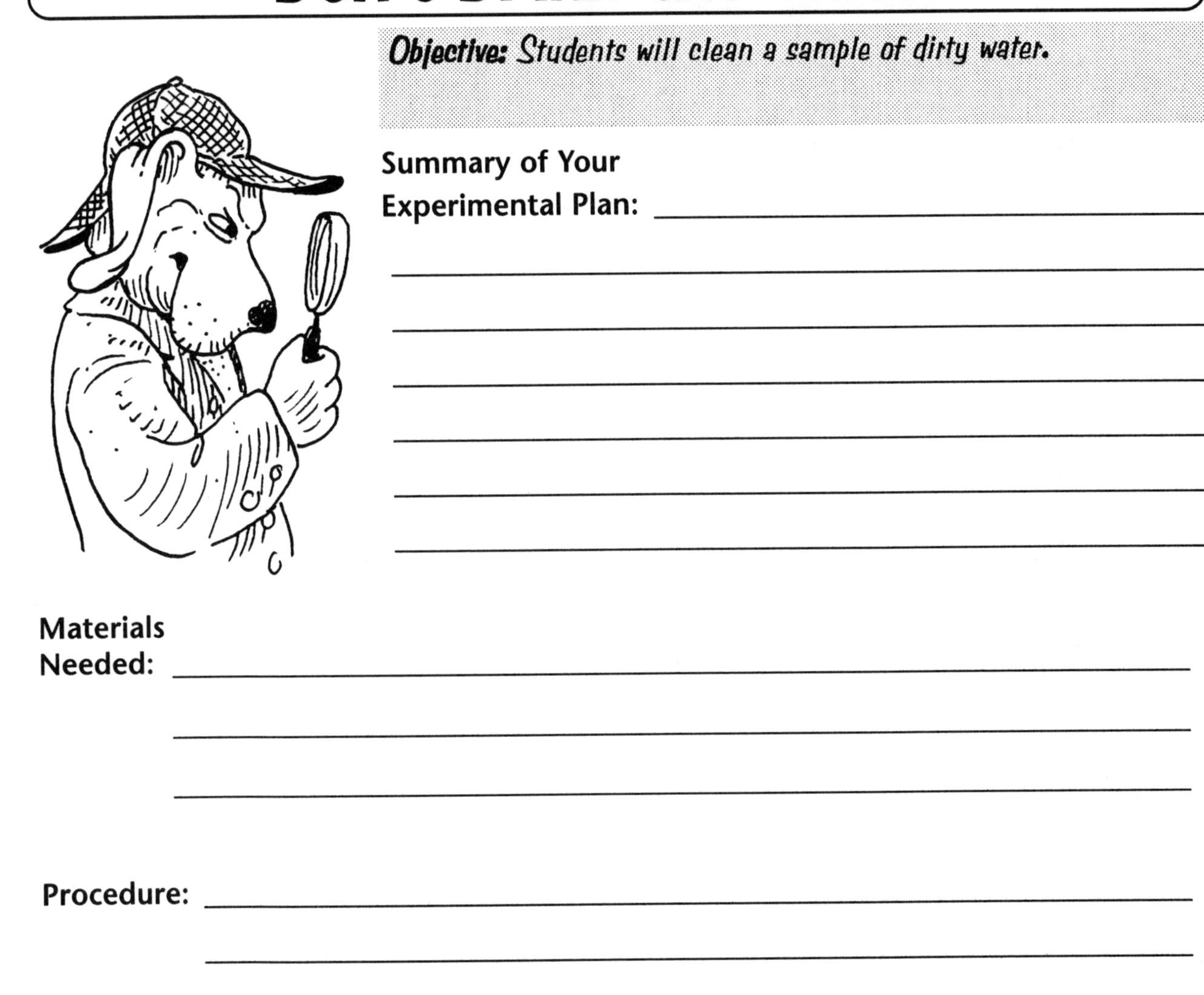

Summary of Your Experimental Plan: ____________________

Materials Needed: ____________________

Procedure: ____________________

Data Table: Quality of Water Before and After Cleaning

Volume	Initial	
	Final	
Clarity	Initial	
	Final	
Odor	Initial	
	Final	
Color	Initial	
	Final	

Results: ____

____

____

____

____

____

____

Conclusion Questions

1. List and describe the water purification steps that you used in this lab.

 ____

 ____

 ____

 ____

2. Which step removed most of the solid matter? ______________________

3. Which step removed most of the odor? ______________________

4. Why do city and county water treatment plants add chlorine to their water? ________

 ____

5. Most city and county water systems do not distill their water, even though distillation removes everything from the water. Why do you think this is so? ____________

 ____

 ____

Strongest Sandstone

Objective: Students will determine how the size of sand particles affects the strength of sandstone.

Time Required
Day 1: 45 minutes
Day 2: 45 minutes

Teaching Strategies

In this partial inquiry, students are provided with Background Information on how sedimentary rock is formed. Students are also given some suggestions about controlling variables in an experiment. However, remind them to use the same amount of glue and water when making all of their sandstone bricks so that these factors will not be variables. A good recipe for a sandstone brick is two ounces of water, one ounce of glue, and five ounces of sand (measured by volume).

Students should select the lab equipment they need from the pieces that you make available to them. Provide them with sand, three sizes of soil sieves, water, water-soluble glue (such as Elmer's), string, weights, paper plates, waxed paper, and aluminum foil.

Suggested Experimental Setup

Using three sizes of soil sieves, sift sand into three containers: large sand particles, medium sand particles, and small sand particles.

Make a graduated measuring cup. Pour one ounce of water in a paper cup, then mark the level with a pen. Add another ounce and mark the level again. Continue in this fashion until you have marked five ounces on the cup.

Mix two ounces of water, one ounce of glue, and five ounces of small sand particles in a bowl. Pour the mixture onto a paper plate; then shape it into a long, rectangular "sandstone rock." Allow to dry.

Repeat this same procedure with medium-sized sand particles and large sand particles.

On Day 2, extend one of the "sandstone rocks" over the edge of the table. Attach a five-gram weight to a string; then hang the string on the "sandstone rock." If it can hold five grams, tie five more grams of weight to the string. Continue adding weight until the "sandstone rock" breaks. Record the amount of weight required to break the "rock."

Repeat this same procedure with the other two "rocks."

Figure 1. Extend a "sandstone rock" over the edge of a table or desk. Attach weights to the rock until it breaks.

Background Information

Strongest Sandstone

? ? ? ? ? Science Sleuth Question ? ? ? ? ?

Which sandstone is the strongest: the type with small sand particles or the kind made of large sand particles?

Rocks are mixtures of minerals that are found beneath the soil and water. That means that no matter what you are doing—studying, swimming, or riding your bike—there is rock underneath you. Some of the soil that can be found on top of rocks was formed in part by their erosion.

Based on how they were formed, rocks can be classified into three categories: igneous, metamorphic, and sedimentary. Igneous rock is formed from hot, melted materials. Sedimentary rock is created from layers of particles that have been pressed together. And metamorphic rock was originally some other kind of rock that was changed by heat or pressure.

Sedimentary rock can be formed in one of three ways:

a. from particles of sand or rock that settle out of water and are pressed together,
b. from the remains of once living animals, or
c. from mineral crystals that were made when the water in which they were dissolved evaporated.

Sandstone is a common sedimentary rock. Depending on the size of the sand particles, sandstone can be weak and crumbly or strong enough to use in construction. In this lab, we will make several samples of "sandstone" by mixing sand with water and glue. To ensure that our results are accurate, all of the "sandstone rocks" must be made by the same recipe, and they should be the same size and shape. Different samples will vary only in the size of the sand particles used to make them.

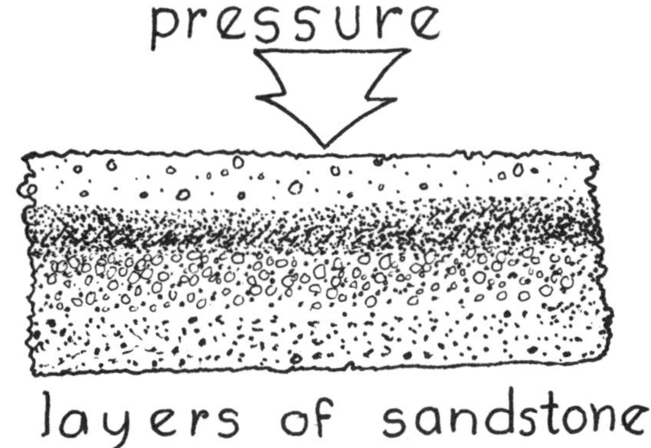

Figure 1. Sandstone is a sedimentary rock that is made when layers of sand particles are deposited one on top of another and then pressed together.

Pre-Lab Questions

1 What are the three types of rocks? ____________________

2. How could an igneous rock be changed into a sedimentary rock? ____________________

____

3. What is the variable that we are testing in this lab? ____________________

____

Name ____________________

Strongest Sandstone

Objective: *Students will determine how the size of sand particles affects the strength of sandstone.*

Summary of Your Experimental Plan: ____________________

____

____

____

____

____

____

Data Table: Strength of Sandstone Rocks Made in Lab

Sand size	Strength (number of grams of weight that the sandstone sample can support when it is extended over the edge of a desk)
Small	
Medium	
Large	

Materials Needed: ____________________

____

____

____

Procedure: ____

____

____

____

____

____

____

____

____

____

____

Results: ____

____

____

____

____

Conclusion Questions

1. In this lab, we tested the strength of your "sandstone rocks" by measuring how much weight the samples could support. Suggest some other ways to measure the strength of the samples. ____

   ____

2. If you were going to use sand to make bricks, what size sand particles would you use?

   ____

   Why? ____

3. Clay is a very small soil particle. Based on your lab work, why do you think that clay is used to make bricks? ____

4. If we used only medium-sized sand particles, what other variable of brick-making could we test? ____

5. Why is it vital to the success of your experiment that you use the same amount of water to make all three sandstone samples? ____

   ____

Minerals on Ice

Objective: Students will determine the freezing point of a 1% salt water solution, a 5% salt water solution, and a 10% salt water solution.

Time Required
Day 1: 55 minutes

Teaching Strategies

In this partial inquiry, students are provided with Background Information about the effect of solvents on freezing point. They should read the entire lab before they begin so they will anticipate the kind of information they are expected to gather.

Consider water as frozen when ice begins to form on the side of the cup or beaker.

For student use in the lab, provide a variety of equipment including water, thermometers, beakers or paper cups, markers, rulers, thermometers, and a freezer or ice chest.

Suggested Experimental Setup

Label three large paper cups as A, B, and C. Have students prepare their salt water samples in the appropriate cups by doing the following:
A. Adding one gram of salt to 100 ml of water to make a 1% solution.
B. Adding five grams of salt to 100 ml of water to make a 5% solution.
C. Adding ten grams of salt to 100 ml of water to make a 10% solution.
Place the three solutions in a cooler of ice or in a freezer. Every five minutes, check the solutions to see whether any freezing has occurred. When ice first begins to form along the edge of the cup, consider the water frozen. Record that temperature in the Data Table.

Figure 1. Make a 1% solution, a 5% solution, and a 10% solution of salt water.

Background Information

Minerals on Ice

? ? ? ? ? Science Sleuth Question ? ? ? ? ?

Can you keep water in the liquid state below its freezing point?

You already know the physical characteristics of pure water. For example, pure water is a colorless, odorless, liquid that freezes at 0° Celsius and boils at 100° Celsius. But what are the characteristics of lake water or ocean water? These surface waters are rarely colorless and odorless. Therefore, it seems unlikely that their boiling and freezing points will be the same as those of pure water.

Figure 1. Pure water freezes at 0° C.

Lake water is described as "fresh" because it does not contain as much salt as ocean water. However, it does contain some dissolved minerals, including salts. The salt with which you are most familiar is sodium chloride, whose formula is NaCl. Sodium chloride is also called *table salt.* Yet there are other salts that are dissolved by water and find their way into lakes and streams.

The ocean contains enough dissolved salt and other minerals to have a salty taste. That is why we describe it as "salt" water. Actually, the percentage of salt and other minerals in ocean water is quite low, about 3.5%. Yet this small amount affects the physical characteristics of ocean water.

In this lab you will compare the physical characteristic of freezing point for several different water samples. The samples will vary in the amount of salt they contain. There are many ways to prepare water samples with different amounts of salt. For example, you could prepare a 1% salt solution by adding one gram of salt to 100 ml of water. Similarly, a 5% salt solution can be made by mixing five grams of salt in 100 ml of water.

Pre-Lab Questions

1. What are two of the physical characteristics of pure water? ______________________

____

2. How could you prepare a 10% salt water solution? A 15% salt water solution? ______

____

3. What kinds of substances are found in lake water? ______________________

4. Why does the ocean taste salty? ______________________

Name ____________________

Minerals on Ice

Objective: *Students will determine the freezing point of a 1% salt water solution, a 5% salt water solution, and a 10% salt water solution.*

Summary of Your Experimental Plan: ____________________

Materials Needed: ____________________

Procedure: ____________________

Data Table: Freezing Points of Salt Water Solutions

Solution	Freezing Point
1% salt water	
5% salt water	
10% salt water	

Results: ____

____

____

____

____

____

____

Conclusion Questions

1. How does the presence of salt in water affect its freezing point? ______________

   ____

2. Salt is sometimes poured on icy walkways and roads. How can salt help prevent freezing? ____

3. To successfully make homemade ice cream, the bucket of ice-cream mix must be constantly turned in a slushy, ice-water solution. Many people add salt to this ice-water slush. What effect would the salt have on the ice-water slush? __________

   ____

   What effect would the salt in the ice-water slush have on the production of ice cream? ____

4. At what temperature would you expect ocean water to freeze? ______________

   Why? ____

5. Would you expect a sample of lake water to freeze at a lower or a higher temperature than ocean water? ____

   Why? ____

6. Graph the information you collected in the Data Table.

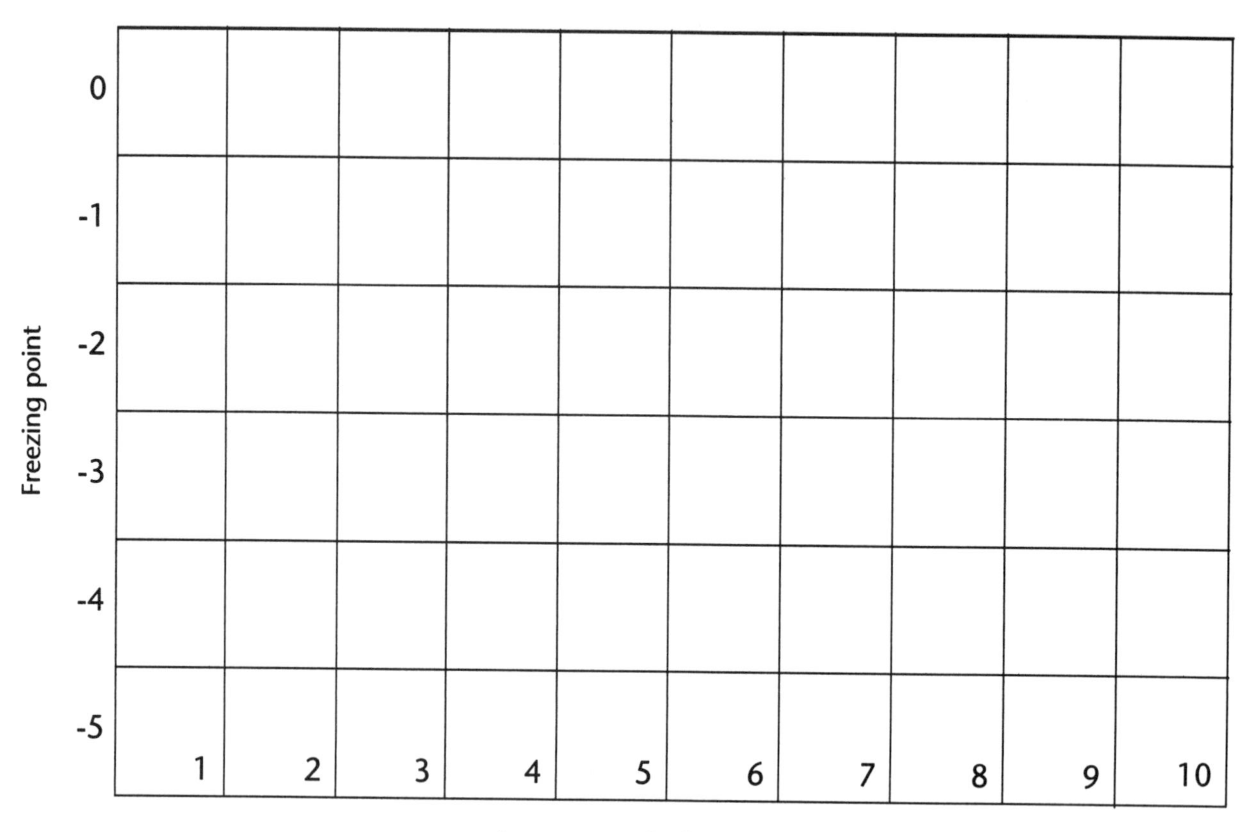

Answer Key

Pre-Lab Questions to Mummified **Page 4**

1. Bacteria and fungi growing on dead plants and animals break them down into simpler substances. This process is known as *decay.*
2. Food must be preserved to keep it from spoiling.
3. a. drying—removing water from dead plants and animals to prevent bacterial growth
   b. salting or sugaring—causing materials to lose water because of high concentrations of salt or sugar
   c. cooling—slowing growth of bacteria and fungi
4. Egyptians bathed the dead in a mixture of chemicals that removed water from the bodies. Without water, bacteria and fungi cannot grow and cause decay. Today we use chemicals that cause changes in tissue, which prevent bacteria and fungi from breaking down the dead tissue.

Conclusion Questions to Mummified **Page 6**

1. Answers will vary but might include apple, grape, or orange.
2. The fruit should appear dry or dehydrated, without any trace of mold. It should not produce a foul odor.
3. It takes several days for the water to be removed from fruit.
4. Mummification of the dead and preservation of food both remove water from the tissues and, therefore, prevent decay.

Pre-Lab Questions to "Lite" Eaters **Page 9**

1. All three terms have similar meanings and usually refer to the absence of some ingredient such as fat.
2. If the liquid is a combination of oil and water, it will separate into two phases. If it is all oil, it will exist in only one phase.
3. Answers may vary, but most students will realize that oil costs more than water.
4. 25-30%
5. Substances like fat and oil that do not dissolve in water.

Conclusion Questions to "Lite" Eaters **Page 11**

1. Answers will vary.
2. Answers will vary.
3. Answers will vary.
4. Answers will vary.
5. Add water to the margarine.

Pre-Lab Questions to Bee Stings and Beef Steaks **Page 13**

1. Proteins can be structural in nature, meaning that they are the material from which living things are made. Other proteins are enzymes, which affect the rate of chemical reactions in the body.
2. Examples of proteins include gelatin, meat, hair, and silk.
3. Enzymes that break down proteins can be found in meat tenderizer, fresh pineapple juice, detergent, and contact lens cleaner.

Conclusion Questions to Bee Stings and Beef Steaks **Page 15**

1. Gelatin
2. The shape changed; the protein lost weight.
3. Student answers will vary; the two enzymes have a similar rate of activity.
4. It takes more than 24 hours for a small quantity of protein-digesting enzyme to completely break down a protein sample.
5. Contact lenses become coated in proteins produced by the eyes.
6. Enzymes in detergents break down protein stains so that they can be washed away.
7. Student answers may vary but might include stomach and intestines.

Pre-Lab Questions to Keep Me Warm and Neutral **Page 18**

1. pH and temperature
2. Enzymes affect the rate of chemical reactions by either holding a substrate so that it can be broken down or so that it can be attached to another piece of substrate to make something new.
3. *Substrates* are substances that are acted upon by enzymes.
4. Answers may vary but could include the breakdown of protein by enzymes.

Conclusion Questions to Keep Me Warm and Neutral **Page 20**

1. Answers may vary but could include warming in a water bath or holding in the sunshine. An increase in temperature reduces enzyme activity.
2. Answers may vary but could include adding vinegar or household ammonia. A change in pH reduces enzyme activity.
3. about 7
4. temperature and pH
5. c. cooler than body temperature
6. You would not be able to digest protein foods.

Pre-Lab Questions to Plants to Dye For **Page 23**

1. dyes made from plants or animal parts
2. Answers may vary but could include madder plant, indigo plant, and saffron crocus.
3. to make the dyed materials colorfast

Conclusion Questions to Plants to Dye For **Page 25**

1. Answers will vary.
2. Answers will vary.
3. Answers will vary.
4. Answers will vary but could include heat the dye and fabric longer, use a different plant to make dye, and use a mordant.
5. No. A mordant was not used in this investigation.

Pre-Lab Questions to Owly's Appetite **Page 27**

1. Hair, feathers, bones, and hard seeds. These materials are not digestible.
2. Owls and hawks cough up their undigested food in pellets.
3. what and how much they ate

Conclusion Questions to Owly's Appetite **Page 29**

1. Answers may vary, but most pellets contain two to five skeletons of small birds or rodents such as voles.
2. Answers may vary but could include something like the following:

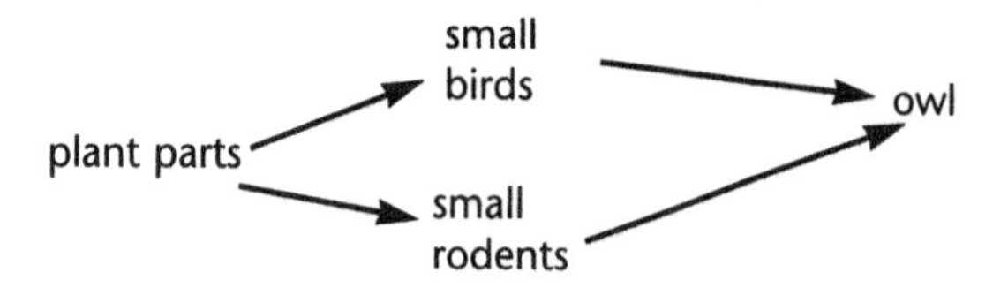

3. Answers may vary. Elimination of owls would alter the food web. There would be fewer rodents and young birds eaten. These two types of organisms could increase their populations to such large degrees that there would not be enough food to support them. Then they would start to die.

True/False Survey to Bacterial Battle **Page 31**

1. T 4. T
2. F 5. T
3. T 6. T

Pre-Lab Questions to Bacterial Battle **Page 31**

1. everywhere
2. heat or disinfectants
3. to slow bacterial growth

Conclusion Questions to Bacterial Battle **Page 33**

1. Answers may vary; probably petri dish C.
2. using clean dishcloths every day, heat, disinfectants
3. slows bacterial growth
4. to prevent cross-contamination

Pre-Lab Questions to Water Bobbers **Page 36**

1. buoyancy
2. The buoyant forces of the water pushing up on the bobber counteract gravity. However, when the mass of the bobber is concentrated into such a small area that little buoyant force can act on it, it sinks.

Conclusion Questions to Water Bobbers **Page 38**

1. salt water
2. ocean water
3. ocean water
4. The boat in the ocean would float higher. The buoyant forces of ocean water are greater than those of fresh water.
5. The upward push of water on an object equals the amount of water displaced.

Pre-Lab Questions to Ink on the Run **Page 40**

1. *Chromatography* is a physical process by which mixtures such as inks are separated into their component colors.
2. Ink is a mixture made of different colored molecules. The variable sizes of the molecules of color determine how far each color is carried by the water.
3. The components of a liquid mixture can be separated by water chromatography.
4. It could not have been separated by water chromatography.

Conclusion Questions to Ink on the Run **Page 42**

1. Answers will vary depending on pens used.
2. No
3. Answers will vary depending on pens used.
4. Investigators could perform chromatography on a section of a check and compare that to chromatography patterns of different pen brands.
5. The green color in leaves can be extracted from the leaves and separated into colored pigments by chromatography.

Pre-Lab Questions to What Is Your Reaction? **Page 45**

1. the formation of a precipitate, color change, production of heat or light, and formation of a gas
2. They are changed into new substances.
3. The original bonds are broken, and new bonds are formed.

Conclusion Questions to What Is Your Reaction? **Page 47**

1. Sugar. Student answers may vary: carbon, water vapor, and carbon dioxide gas were formed.
2. color change
3. NaCl has stronger bonds. Its bonds do not break with the application of heat from a Bunsen burner or hot plate.
4. Yes. Bubbling occurs in the soda bottle and the balloon is inflated by a gas.

Pre-Lab Questions to Plop, Plop, Fizz, Fizz **Page 50**

1. Physical change—Answers may vary but could include breaking the egg.
   Chemical change—Answers may vary but could include cooking the egg.
2. They increase the surface area of the onion and, therefore, increase the onion flavor in the soup.
3. In a chemical change, reactants are changed into different products. In a physical change, new products are not formed.
4. The clay's surface area is that part of the clay exposed to the air—the outside surface. You could increase the surface area by cutting, chopping, or mashing the clay.
5. b.
   d.
   e.

Conclusion Questions to Plop, Plop, Fizz, Fizz — Page 52

1. the reaction of an antacid tablet with water
2. water and antacid tablets
3. Fizzing occurs during the chemical reaction.
4. the pile of wooden splinters—They have more surface area than the log.
5. the crushed piece—It has more surface area.

Pre-Lab Questions to Putting on the Heat — Page 54

1. variations in pressure, temperature, concentration, and surface area
2. The molecules must collide with enough energy to break bonds and reform them.
3. Cutting a potato into strips increases its surface area.

Conclusion Questions to Putting on the Heat — Page 56

1. Answers will vary, depending on process used. Most will use a stopwatch to determine how long the effervescent antacid tablet fizzes.
2. It increases the speed of the reaction.
3. The cold air slows the decay of the body.
4. warm water. The warm water increases the rate of the reaction and the process of respiration. Respiration occurs slowly in cool water.

Pre-Lab Questions to Ice Sculpture — Page 60

1. A *glacier* is a large mass of ice that moves slowly across the land.
2. Glaciers erode the soil and transport materials.
3. They are rounded by the tumbling they do as they travel with the glacier.

Conclusion Questions to Ice Sculpture — Page 63

1. The ice reduced the height of the mound of soil.
2. Student answers will vary.
3. The ice increased the width and length of the mound of soil.
4. Yes. Answers on height will vary, depending on the mass of the ice cubes and the type of soil.
5. Answers will vary. Moraines and gorged-out valleys are evidence of the work of glaciers.

Pre-Lab Questions to Soil on the Move — Page 65

1. a, b, c, d, e, g, h
2. *Erosion* is the wearing away of soil by wind, water, and ice.
3. Farmers plow furrows around the hill rather than up and down the slope. Students might mention that farmers leave some plant cover over the soil.

Conclusion Questions to Soil on the Move — Page 67

1. It increases.
2. It increases. Answers may vary somewhat, but students should explain that fast-moving water can push or pick up more mass than water that is moving slowly.
3. The force of flowing water can push dirt. It can also carry it as suspended and/or dissolved matter.
4. Answers will vary, but might include amount of soil, amount of water, temperature of water, and type of soil.
5. Agree. The variable that we are testing in this lab is the speed of water (which is due to the slope).

Pre-Lab Questions to The Frozen Explosion — Page 70

1. solid, liquid, and gas
2. a. liquid
   b. gas
   c. gas
3. Water enters rocks as a liquid, freezes and expands, breaking the rocks in the process.
4. Polar molecules have a positive end and a negative end. Water is a polar molecule.

Conclusion Questions to The Frozen Explosion — Page 72

1. The water will freeze, expand, and burst the can because water in a solid form takes up more space than it does in liquid form.
2. The rocks break.
3. High in the Rocky Mountains where the weather is warm in summer and cold in winter. Water changes from the liquid to the solid state more often in the Rockies than it does in the tundra.
4. Water is a polar molecule. As molecules cool, they slow down and arrange themselves so that oppositely charged ends are close together. This arrangement takes up more space than the fast-moving molecules of liquid water.
5. Freezing water in cells expands and breaks the cell membranes, causing damage.
6. Data tables will vary depending on student's original volume.

Pre-Lab Questions to Don't Drink the Water — Page 76

1. Drinking water contains many pollutants such as litter, oil, and bacteria.
2. a. settling
   b. filtration
   c. adsorption
   d. distillation
   e. disinfection
3. Alum attracts clay and other suspended matter and helps it settle to the bottom.

Conclusion Questions to Don't Drink the Water — Page 79

1. Answers will vary.
2. Answers will vary.
3. Answers will vary.
4. to kill disease-causing bacteria
5. Answers will vary; students might mention that distillation removes everything, including minerals that give flavor, or they might express some concern that distillation is expensive.

Pre-Lab Questions to Strongest Sandstone Page 81

1. igneous, metamorphic, and sedimentary
2. Erosion could change igneous rock to small particles that could be deposited and pressed together.
3. the size of the sand particles

Conclusion Questions to Strongest Sandstone Page 83

1. Student answers will vary but might include dropping the samples from a standard height or pulling on them with a spring scale.
2. small—Students will probably find that their sandstone bricks made of small sand particles are the strongest because smaller particles are pressed more closely together and, consequently, produce stronger bricks.
3. Small-sized soil particles make strong bricks and rocks.
4. Student answers will vary but might include amount of water or amount of glue.
5. It is only possible to test one variable at a time. If two variables exist in an experiment, the results are difficult to interpret.

Pre-Lab Questions to Minerals on Ice Page 85

1. odorless and colorless liquid, boiling point 100° C and freezing point 0° C
2. Prepare a 10% salt water solution by adding 10 grams of salt to 100 ml of water; make a 15% solution by adding 15 grams of salt to 100 ml of water.
3. Student answers may vary but might include dissolved minerals, suspended soil, and dead plant and animal matter.
4. It contains sodium chloride and other salts.

Conclusion Questions to Minerals on Ice Page 82

1. Salt lowers the freezing point of water.
2. Salty water can exist as a liquid at colder temperatures than pure water. Therefore, salt can prevent cold water from turning to ice.
3. The addition of salt allows the water to remain a liquid at a temperature below 0° C. Colder water in the freezer helps the ice cream freeze faster.
4. Student answers will vary but should indicate a temperature below zero degrees Celsius since the salt in the ocean water lowers its freezing point.
5. Higher. Lake water does not contain as much salt (or dissolved minerals) as ocean water.
6. Data tables will vary depending on the volume students used.